When Life is Linear

From Computer Graphics to Bracketology

© 2015 by The Mathematical Association of America (Incorporated)

Library of Congress Control Number: 2014959438

Print edition ISBN: 978-0-88385-649-9

Electronic edition ISBN: 978-0-88385-988-9

Printed in the United States of America

Current Printing (last digit):
10 9 8 7 6 5 4 3

When Life is Linear

From Computer Graphics to Bracketology

Tim Chartier

Davidson College

Published and Distributed by

The Mathematical Association of America

To my parents, Jan and Myron,
thank you for your support,
commitment, and sacrifice
in the many nonlinear
stages of my life

MAA Service Center
P.O. Box 91112
Washington, DC 20090-1112
1-800-331-1MAA   FAX: 1-240-396-5647

Contents

Preface

In the 1999 film *The Matrix*, there is the following interchange between Neo and Trinity.

> Neo: *What is the Matrix?*
> Trinity: *The answer is out there, Neo, and it's looking*
>   *for you, and it will find you if you want it to.*

As an applied mathematician specializing in linear algebra, I see many applications of linear systems in the world. From computer graphics to data analytics, linear algebra is useful and powerful. So, to me, a matrix, connected to an application, is indeed out there–waiting, in a certain sense, to be formed. When you find it, you can create compelling computer graphics or gain insight on real world phenomenon.

This book is meant to engage high schoolers through professors of mathematics in applications of linear algebra. The book teaches enough linear algebra to dive into applications. If someone wanted to know more, a Google search will turn up a lot of information. If more information is desired, a course or textbook in linear algebra would be the next step. This book can either serve as a fun way to step into linear algebra or as a supplementary resource for a class.

My hope is that this book will ignite ideas and fuel innovation in the reader. From trying other datasets for data analytics to variations on the ideas in computer graphics, there is plenty of room for your personal discovery. So, this book is a stepping stone, intended to springboard you into exploring your own ideas.

This book is connected to a Massive Open Online Course (also known as a MOOC) that will launch in February of 2015 through Davidson College and edX. The book and online course are not dependent on each other. Participants of the MOOC do not need this book, and readers of this book do not need to participate in the MOOC. The experiences are distinct, but for those interested, I believe the book and online course will complement each other. One benefit of the MOOC is a collection of online resources being developed. Many of the ideas in this book require a computer and either mathematical software or a computer program to calculate the underlying mathematics. From graphics that have thousands of pixels to datasets that have hundreds or thousands of entries, computing can be necessary even for ideas that can be easily demonstrated by hand on small datasets. I look forward to seeing the ideas of this book shared, explored, and extended with the online community.

The book has one more goal. A few years ago, a group of students gathered for a recreational game of Math Jeopardy. One of the subject areas was linear algebra, which turned out to be the hardest for the students. As the students struggled to recall the content necessary to answer the questions, one student joked, "Well, at least this stuff isn't useful. Maybe that's why I can't remember it." Even looking at the table of contents of this book, you'll see why I see linear algebra as very useful. I tend to believe that at any second in any day linear algebra is being used somewhere in a real world application. My hope is that after reading this book more people will say, "I really need to remember these ideas. They are so useful!" If that person is you, maybe you'd begin sharing the many ideas you created and explored as you read this book. I wish you well as you discover matrices awaiting to be formed.

Acknowledgments

First and foremost, I'd like to thank Tanya Chartier, who travels this journey of life with me as my closest companion. Our paths, while not always linear, have always felt parallel. Thanks to Karen Saxe, chair of the Anneli Lax New Mathematics Library Editorial Board, for her encouragement, many insights, and commitment to this project. Along with Karen, I thank the Anneli Lax New Mathematics Library Editorial Board for their reading of the book and insightful edits that improved the book and its content. I want to thank Michael Pearson, Executive Director of the Mathematical Association of America, for his support for this project and its connection to the MOOC. I'd also like to thank Shane Macnarmara for diving into the development of the online resources that accompany the MOOC and, as such, this book as well. Shane's commitment enhanced the online resources and will benefit many people who use them. Finally, I want to thank the folks involved in the Davidson College MOOC, Allison Dulin, Kristen Eshleman, and Robert McSwain, for the many ways their work on the online course enhanced the content in these printed pages. I am grateful to my colleagues, near and far, who support my work and to the many students who sit and chat about such things and offer insightful ideas that enhance and enrich my activities. It is impossible to thank everyone who is involved in a project and supports its content. For those unmentioned, may you see your influence in these pages and enjoy the way our interaction materialized into printed work.

1
X Marks the Spot

In this book, we will think linearly. In two dimensions, this means a line. In three dimensions, we're talking about a plane. In higher dimensions, math helps us work with hyperplanes. A lot of the world isn't linear. The world, especially the natural world, often offers beautiful curves. Yet, like the horizon, curvature, if viewed in the right way, can look linear.

The ability to approximate curves with lines will be important to many portions of this book. To get a visual sense of modeling curved space with lines, consider sketching only with straight lines. How about drawing a portrait? I'll lay down a series of dots that approximate an image and then I'll draw one continuous line through all the points. I'll start and end at the same point. See Figure 1.1 for an example. Recognize the image in Figure 1.1? Such visual art, called TSP Art, was introduced and developed in [3, 4]. "TSP" stands for "traveling salesperson problem" since the underlying dots can be viewed as cities and the line segments between dots indicate the route the salesperson will make through the cities. Such problems help minimize travel.

Though the image is not the original portrait, the drawing is recognizable. The line drawing captures, in this case, visual components of the original image. Later, we will use linear phenomenon to model sports performance enabling us to predict future play.

But, let's not get too far ahead of ourselves. Most of this book will explore linear systems, essentially puzzles written as equations. Let's see an example that I'll pose in the form of a magic trick.

Figure 1.1. A portrait of Marilyn Monroe drawn with straight lines. Photograph courtesy of Robert Bosch.

Think of a number between 1 and 20. Double the number. Now, add 8. Next, divide by 2. Subtract your original number and mark the spot on the number line where this computed number lies with an x.

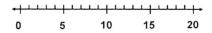

I bet I know where x lies on the number line. It lies at the value 4. How did I know? I did this by first using a variable y, and y equals the original number you chose, regardless of its value. In the second step, you doubled it producing $2y$. Adding 8 resulted in $2y + 8$, and then dividing by 2 led to $(2y + 8)/2 = y + 4$. Finally, I asked you to subtract that original number. I never knew its value but did know that it equalled y so, when you subtract y from $y + 4$, you indeed get 4. So, my guess wasn't much of a guess; rather it was some algebra—in fact, *linear* algebra since all the terms (y, $2y$, $2y + 8$ and so forth) are linear. I never needed to know your original number. In fact, it wouldn't have mattered if you selected a number between -100 and 100 or a number with a million digits. Regardless, your final value would always result in 4.

Solving for unknowns is a powerful tool of linear algebra. Here's a problem you may recall from a math class that again looks for one unknown number.

The Alpha Train, traveling 80 miles per hour (mph), leaves Westville heading toward Eastburg, 290 miles away. Simultaneously,

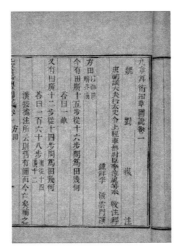

Figure 1.2. The *Jiuzhang Suanshu*, a Chinese manuscript dating from approximately 200 BC. Pictured is the opening of Chapter 1.

the Beta Train, leaves Eastburg, traveling 65 mph, toward Westville. When do the two trains meet?

Here x will again denote the unknown. In this case, we translate the word problem; $80x + 65x = 290$. With that, we solve and find $x = 2$. In two hours, the trains will meet when the total of their distances equals the distance between the cities.

Puzzles like this date back thousands of years. Here's one from a Chinese manuscript from approximately 200 BC called the *Jiuzhang Suanshu*, which contained 246 problems intended to illustrate methods of solution for everyday problems in areas such as engineering, surveying, and trade. [12]

> There are three classes of grain, of which three bundles of the first class, two of the second, and one of the third make 39 measures. Two of the first, three of the second, and one of the third make 34 measures. And one of the first, two of the second, and three of the third make 26 measures. How many measures of grain are contained in one bundle of each class?

This problem involves three unknowns rather than one as in our problem with the train.

Let x denote the number of measures of the first class grain in one bundle. Let y and z equal the number of measures of the second and third class grains in one bundle, respectively. The word problem can be written mathematically as

$$
\begin{aligned}
3x + 2y + z &= 39 \\
2x + 3y + z &= 34 \\
x + 2y + 3z &= 26.
\end{aligned}
$$

Can you solve the puzzle from over twenty centuries ago? Give it a try. When you're ready, look at the end of this chapter for the answer.

In this book, we will see that systems of linear equations have many applications. So far, the systems of linear equations we've seen have had one equation and one unknown (with the train) and three equations and three unknowns (with the grains). Google uses linear systems with billions of unknowns to help with its search engine.

Science, engineering, business, government and even Hollywood use systems of linear equations to aid in their applications. In the coming chapters, we'll explore applications of linear algebra ranging from finding a presidential look-alike to recommending a movie.

Regarding the problem from the *Jiuzhang Suanshu*, it led to the linear system

$$
\begin{aligned}
3x + 2y + z &= 39 \\
2x + 3y + z &= 34 \\
x + 2y + 3z &= 26,
\end{aligned}
$$

which leads to the solution $x = 9.25$, $y = 4.25$ and $z = 2.75$. We now know how many measures of grain are contained in one bundle of each class.

2

Entering the Matrix

In this section, we explore matrices, mathematical structures of great use. By the time you've turned the last page of this book, we will have rotated a 3D computer image, learned sports ranking methods, and delved into a mathematical algorithm that helped build the foundation of Google. So if you are wanting to find an application of a matrix, you won't need to look long or hard. The applications are many, awaiting your exploration.

Matrices can be thought of as tables, and soon we'll be applying mathematical operations to them. For us, the power of a matrix comes through its properties as a mathematical object and not just as a structure for storage. Let's begin with the anatomy of a matrix by defining

$$A = \begin{bmatrix} 3 & 0 & 3 \\ 8 & 8 & 8 \end{bmatrix}.$$

This matrix A has two rows and three columns, which is often denoted as having size 2×3. The (i, j)th element is found in the ith row and jth column. So, element $(1, 2)$ in matrix A, also denoted $a_{1,2}$, equals 0.

In many applications, the digits contained in a matrix connect to an application. That's true here too. In fact, the numbers in A are part of a much larger matrix B, seen by the shading in (2.1), in which

$$B = \begin{bmatrix} 1 & 1 & 1 & 1 & 1 & 1 & 1 & 1 & 1 & 1 & 1 & 1 & 1 & 1 & 1 \\ 0 & 0 & 1 & 1 & 0 & 0 & 3 & 0 & 3 & 3 & 3 & 3 & 0 & 0 & 3 \\ 0 & 8 & 5 & 8 & 8 & 5 & 8 & 8 & 8 & 8 & 8 & 5 & 5 & 5 & 5 \\ 8 & 8 & 8 & 3 & 0 & 3 & 0 & 3 & 3 & 8 & 5 & 8 & 8 & 5 & 8 \\ 3 & 3 & 0 & 0 & 1 & 1 & 0 & 0 & 3 & 8 & 5 & 8 & 5 & 5 & 5 \end{bmatrix}. \tag{2.1}$$

Mathematically, A is called a *submatrix* of B. Do the elements seem random? They aren't. But again, to see why, we need to see the matrix for which B

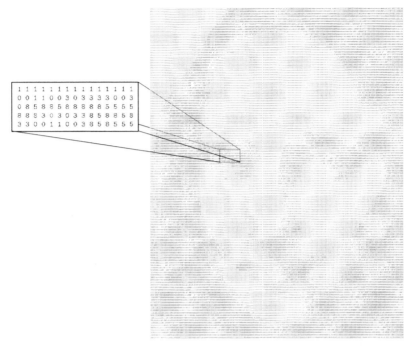

Figure 2.1. A well-known image created with a printer and numbers, but to much less satisfying effect than with a brush and paint.

is a submatrix, which you'll see in Figure 2.1. So, where is A? It's part of Mona Lisa's right eye.

Here lies an important lesson in linear algebra. Sometimes, we need an entire matrix, even if it is large, to get insight about the underlying application. Other times, we look at submatrices for our insight. At times, the hardest part is determining what part of information is most helpful.

2.1 Sub Swapping

As we gain more tools to mathematically manipulate matrices, we'll look at applications such as finding a group of high school friends among a larger number of Facebook friends. We'll identify such a group with submatrices. The tools that we develop will help with such identification. While such mathematical methods are helpful, we can do interesting things simply with submatrices, particularly in computer graphics.

(a) (b)

(c)

Figure 2.2. Images of a lion cub (a), adult lion (b), and composite images (c). Photo of lion cub by Sharon Sipple, (www.flickr.com/photos/ssipple/14685102481/), and photo of adult lion Steve Martin, (www.flickr.com/photos/pokerbrit/14154272817/), CC BY 2.0 (creativecommons.org/licenses/by/2.0).

To see this, let's take the image of lion cub and adult lion seen in Figure 2.2 (a) and (b). Can you figure out how I produced the images in Figure 2.2 (c)? Take a moment and consider it before reading on.

The images in Figure 2.2 (c) were created by randomly choosing a pixel location in the image of the lion cub. Then, a random size of a submatrix was chosen. Then each of the submatrices of the chosen size with its upper righthand corner at the chosen pixel location were swapped between the images of the cub and adult lion. Doing this several times formed the composite images in Figure 2.2 (c).

While we are clearly choosing small rectangular portions of a larger image, how does this connect to matrices? Where are the numbers? In Figure 2.1, the numbers formed the image. In this example, the numbers are equally important and, in their own way, play a role in what we see. To see this, let's make everything black and white, or actually gray.

2.2 Spying on the Matrix

Any picture can be viewed as a matrix, and matrices can be viewed as images to gain insight on the data. Let's start with the numbers by looking at the matrix

$$M = \begin{bmatrix} 128 & 126 & 97 & 100 & 99 & 100 & 99 & 118 & 138 & 130 & 122 & 135 \\ 64 & 48 & 39 & 42 & 49 & 45 & 59 & 71 & 81 & 93 & 90 & 118 \\ 26 & 27 & 29 & 42 & 58 & 34 & 34 & 25 & 20 & 29 & 46 & 67 \\ 36 & 38 & 45 & 34 & 20 & 22 & 19 & 53 & 40 & 20 & 28 & 46 \\ 59 & 58 & 48 & 51 & 33 & 21 & 24 & 23 & 45 & 33 & 42 & 34 \\ 71 & 66 & 48 & 47 & 40 & 25 & 26 & 23 & 54 & 41 & 48 & 55 \\ 70 & 59 & 53 & 56 & 38 & 36 & 37 & 29 & 22 & 11 & 37 & 65 \\ 68 & 57 & 66 & 41 & 43 & 31 & 32 & 26 & 33 & 41 & 63 & 67 \\ 53 & 47 & 90 & 77 & 80 & 43 & 42 & 26 & 44 & 79 & 89 & 89 \\ 77 & 44 & 61 & 91 & 82 & 91 & 76 & 63 & 74 & 98 & 92 & 71 \\ 116 & 61 & 42 & 55 & 67 & 81 & 86 & 77 & 82 & 103 & 84 & 47 \\ 118 & 117 & 50 & 27 & 35 & 44 & 54 & 68 & 88 & 93 & 80 & 83 \\ 150 & 138 & 134 & 112 & 103 & 142 & 147 & 144 & 151 & 147 & 141 & 115 \\ 168 & 172 & 187 & 166 & 177 & 179 & 162 & 160 & 142 & 133 & 131 & 139 \\ 170 & 170 & 178 & 159 & 169 & 153 & 135 & 127 & 133 & 131 & 118 & 119 \\ 154 & 150 & 145 & 131 & 135 & 126 & 133 & 125 & 114 & 108 & 104 & 115 \end{bmatrix}.$$

This matrix M stores the data of a grayscale image. Since the matrix has sixteen rows and twelve columns, the associated image would have sixteen rows and twelve columns of pixels. All the values are between 0 and 255 where 0 stores black and 255 white. So, the upper left-hand pixel has a value of 128 and would be gray. A zoomed-in view of the corresponding image is seen in Figure 2.3 (a). Like the previous section, we again have a submatrix, in this case of the matrix corresponding to the image in Figure 2.3 (b).

Even when the data in a matrix isn't related to a picture, graphically viewing the table of numbers can be helpful. One application is finding structure in a matrix by creating a plot of the matrix, sometimes called a

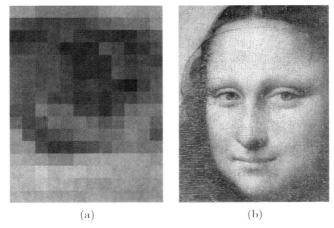

(a)   (b)

Figure 2.3. A detail (a) of the Mona Lisa (b). Can you eye (that's a hint) the location of the detail in (a) in the larger image in (b)?

spy plot. Let's work with the matrix

$$
S = \begin{bmatrix}
0 & 0 & 0 & 0 & 0 & 0 & 0 & 0 & 0 & 0 \\
0 & 0 & 0 & 0 & 0 & 0 & 1 & 1 & 1 & 0 \\
0 & 0 & 0 & 0 & 0 & 0 & 1 & 1 & 1 & 0 \\
0 & 0 & 0 & 0 & 0 & 0 & 1 & 1 & 1 & 0 \\
0 & 0 & 0 & 0 & 0 & 0 & 0 & 0 & 0 & 0 \\
0 & 0 & 0 & 0 & 0 & 0 & 0 & 0 & 0 & 0 \\
0 & 1 & 1 & 1 & 0 & 0 & 0 & 0 & 0 & 0 \\
0 & 1 & 1 & 1 & 0 & 0 & 0 & 0 & 0 & 0 \\
0 & 1 & 1 & 1 & 0 & 0 & 0 & 0 & 0 & 0 \\
0 & 0 & 0 & 0 & 0 & 0 & 0 & 0 & 0 & 0
\end{bmatrix}.
$$

When we create a spy plot, we put a white square in the place of every zero element and fill it with black otherwise. Creating a spy plot for S produces the image in Figure 2.4 (a). Take a moment to ensure you see the pattern in the elements of S as they appear in the spy plot. When you have, what matrix would make the spy plot in Figure 2.5 (a)? Then, consider the optical illusion in Figure 2.5 (b).

Recognizing structure in a matrix can be helpful. Later in the book, we'll learn to solve systems of linear equations, which aids in finding solutions like the problem posed in the *Jiuzhang Suanshu* in Chapter 1. For some problems,

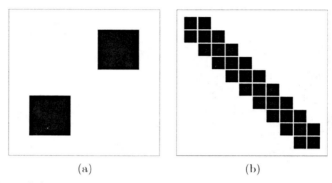

Figure 2.4. A simple matrix pattern (a) and a banded matrix's pattern (b).

the time to solution can be greatly reduced if the associated matrix has a spy plot of the form seen in Figure 2.4 (b). The reduction in time is more striking as the matrix grows, such as a matrix with 1,000 rows rather than 10 as pictured. What do you notice about the structure of the matrix in Figure 2.4 (b)? Such matrices are called *banded*.

2.3 Math in the Matrix

Before moving to the next section and performing mathematical operations on matrices, let's learn to construct a matrix that, when viewed from afar, creates an image like that of the Mona Lisa that we saw in Figure 2.1.

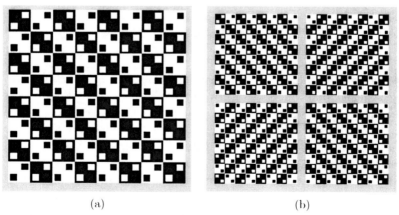

Figure 2.5. Using the pattern in Figure 2.4 (a) to create optical illusions.

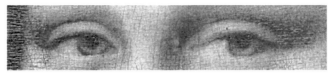

Figure 2.6. Detail of the Mona Lisa that will be approximated with the numbers in a matrix.

Let's work a bit smaller and create a matrix of Mona Lisa's eyes as seen in Figure 2.6.

We will break the interval from 0 to 255 into four intervals from 0 to 63, 64 to 127, 128 to 191, and 192 to 255. All pixels from 0 to 63 will be replaced by a 5 or an 8. All pixels from 64 to 127 will be replaced by a 0 or a 3. All remaining pixels will be replaced by a 1. When there is more than one choice for a number, like choosing a 5 or an 8, we choose a value randomly with a flip of a coin.

The darkest pixels are replaced with a 5 or an 8 as they require more pixels to draw than a 1, which replaces the lightest pixels. Performing this replacement process on the Mona Lisa's eyes creates the image seen in Figure 2.7. If you look carefully at it, you may notice that Mona Lisa's face is thinner. The aspect ratio is off. This is due to the fact that typed characters, particularly in this font, do not fill a square space, but instead a rectangular one, which can be seen in the text below.

DON'T BE SQUARE

Each character is almost twice as tall as it is wide. So, we will adapt our algorithm.

Figure 2.7. Image of the Mona Lisa that replaces each pixel with a number.

Figure 2.8. Image of the Mona Lisa that replaces pixels with a number.

To keep the aspect ratio of the original image, we'll replace two pixels with a number, working specifically with 2×1 submatrices of the larger matrix, which should have an even number of rows. Let's see the steps on the matrix

$$\begin{bmatrix} 71 & 192 & 215 & 90 & 20 \\ 174 & 66 & 65 & 212 & 14 \\ 168 & 130 & 208 & 150 & 136 \\ 42 & 179 & 63 & 141 & 199 \\ 31 & 228 & 237 & 234 & 239 \\ 128 & 245 & 90 & 73 & 34 \end{bmatrix}.$$

We start at the upper lefthand corner of the matrix and partition it into non-overlapping 2 by 1 submatrices

$$\begin{bmatrix} 71 & 192 & 215 & 90 & 20 \\ 174 & 66 & 65 & 212 & 14 \\ 168 & 130 & 208 & 150 & 136 \\ 42 & 179 & 63 & 141 & 199 \\ 31 & 228 & 237 & 234 & 239 \\ 128 & 245 & 90 & 73 & 34 \end{bmatrix}.$$

We then replace each submatrix with a single element equalling the average of the elements in it. For our example, this becomes

$$\begin{bmatrix} 122.5 & 129 & 140 & 151 & 17 \\ 105 & 154.5 & 135.5 & 145.5 & 167.5 \\ 79.5 & 236 & 163.5 & 153.5 & 136.5 \end{bmatrix}.$$

Note that the $(1,1)$ element equals 122.5, which is the average of 71 and 174.

Finally, we replace each element with the corresponding number as we did earlier. So, 122.5 would be replaced with a 5 or 8. Performing this algorithm on the image in Figure 2.6, the image in Figure 2.8 is formed. When performed on the larger image of the Mona Lisa, we get the image in Figure 2.1.

In this chapter, we've seen applications of a matrix. We are now ready to apply mathematical operations and see applications related to them.

3
Sum Matrices

Now, we are ready to perform arithmetic with the matrices that we learned to build in the last chapter. It will take only a few pages to learn how to perform arithmetic on matrices. If you scan forward, there are a number of pages that follow in this chapter. This offers an important lesson about mathematics. Sometimes, real-world applications require complicated mathematical methods. But, it is quite important to realize that simple tools can lead to interesting applications, as we'll see in this chapter. In fact, sometimes, the difficulty in applying math isn't in the complexity of the mathematical method but more in recognizing that a method, which can be quite simple, *can* be applied.

3.1 Adding to Things

Adding matrices is easy enough. First, both matrices must have the same number of rows and columns. Then, we add element-wise, which is most easily seen in an example. Let

$$A = \begin{bmatrix} 1 & 2 & 3 \\ 6 & -1 & 4 \end{bmatrix} \quad \text{and} \quad B = \begin{bmatrix} 0 & 4 & 8 \\ -5 & 1 & 2 \end{bmatrix}.$$

Then

$$A + B = \begin{bmatrix} 1+0 & 2+4 & 3+8 \\ 6+(-5) & -1+1 & 4+2 \end{bmatrix} = \begin{bmatrix} 1 & 6 & 11 \\ 1 & 0 & 6 \end{bmatrix}.$$

The other arithmetic operation we'll learn is scalar multiplication, which multiplies a matrix by a real number or as we'll often call it, a *scalar*. So, for example, if $B = 3A$ then A and B are the same size and each

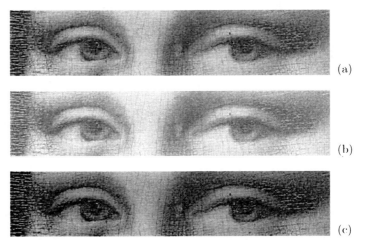

(a)

(b)

(c)

Figure 3.1. Detail of the Mona Lisa (a) lightened (b) darkened (c) with matrix arithmetic.

element of B equals the corresponding element of A multiplied by 3. So,

$$3\begin{bmatrix} -1 & 4 & 0 \\ 3 & -3 & 2 \end{bmatrix} = \begin{bmatrix} -3 & 12 & 0 \\ 9 & -9 & 6 \end{bmatrix}.$$

Like multiplication with scalars, multiplication by a positive integer can be viewed as repeated addition. For example, $3A = A + A + A$. So, in a way, we've simply learned to add matrices in this chapter whether through matrix addition or scalar multiplication. Armed with these matrix operations, we are ready to delve into applications.

To get our applied math ball rolling, let's use matrix addition on the image that closed the last chapter, the eyes of the Mona Lisa as seen in 3.1 (a). Let M be the matrix containing the grayscale information of this image. By simply adding 40 to every element of M, we lighten the image as seen in Figure 3.1 (b). What happens if we subtract 40? The result is in Figure 3.1 (c). Note, subtracting 40 from every element can result in negative values. Such pixels are colored black. Similarly, values exceeding 255, which can occur in addition, are treated as white.

A strong advantage of matrix addition is that its simplicity also leads to its efficiency. Matrix addition can be implemented efficiently on computers allowing immediate response to user input when used in applications like computer games.

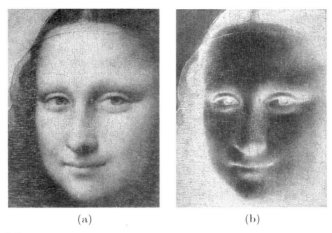

(a)   (b)

Figure 3.2. The grayscale intensities of the original image (a) are inverted to create the image in (b).

3.2 Getting Inverted

Now, let's apply scalar multiplication and matrix addition again on the grayscale image of the Mona Lisa. Specifically, let's invert the colors of the image in Figure 3.2 (a).

Think about how this would be done with one pixel. If such a pixel were black, its grayscale value would be 0.

We want to change the pixel to white, that is, to the value 255. Similarly, if the pixel were white (a value of 255), the new grayscale intensity should be 0. Think of the line passing through the two points (255, 0) and (0, 255), which has equation

$$y = -x + 255.$$

How do we scale this up to a matrix that contains the intensities of all the pixels? Let M equal the matrix of grayscale values that form the original image in Figure 3.2 (a) and N the inverted image in Figure 3.2 (b). To form the new image with inverted colors, take

$$N = -M + 255U,$$

where U is a matrix (the same size as M) with all 1s. Precisely this mathematical manipulation created the image in Figure 3.2 (b).

3.3 Blending Space

Inspired by a historic speech by President John F. Kennedy, let's use linear algebra to visually capture a moment in time that committed the United States to a national space program. In the speech, the president stated, "We go into space because whatever mankind must undertake, free men must fully share." Soon after, he made the resounding challenge, "I believe that this nation should commit itself to achieving the goal, before this decade is out, of landing a man on the moon and returning him safely to the Earth." By the end of the decade, what might have seemed like a distant dream in his speech became reality with the Apollo 11 mission to the moon.

We start with images of President Kennedy giving that address before a joint session of Congress in 1961 and Buzz Aldrin standing on the moon as seen in Figures 3.3 (a) and (b). Using Photoshop, I cut the astronaut out of Figure 3.3 (b) and placed him on a black background as seen in Figure 3.3 (c). My goal is to blend the two images into a new image. The choice of a black background in Figure 3.3 (c) is due to the grayscale value of black pixels, which is zero. In a moment, we'll discuss the implication of a black background more.

Let J and B denote the matrices containing the grayscale information of JFK and Buzz Aldrin, respectively. To create the blend, we let $A = \alpha J + (1 - \alpha)B$, where A becomes the blended image. If $\alpha = 0.5$ then A has equal contributions from each picture. I used $\alpha = 0.7$ to create Figure 3.3 (d), which created each pixel with 70% of the value from image of JFK and 30% from the image of the astronaut. This resulted in the image of JFK being more defined and prominent than that of the astronaut.

Let's return a moment to the clipped image of Buzz Aldrin in Figure 3.3 (c) as opposed to (b). Had I used the unclipped image, the surface of the moon would have been visible, albeit faintly, in the blended image. Not including this was my artistic choice. Let's also look at the decision of having a black background, rather than using another color like white. Again, a black grayscale pixel has a value of 0. If p is the value of a pixel in the image of JFK and the pixel in the same location is black in clipped image of Buzz Aldrin, then the value of the pixel in the blended image is αp, which equals $0.7p$. We know this results in a slightly darker image of JFK giving his speech. Suppose I had placed Aldrin on a white background. What result would you anticipate?

Figure 3.3. Exploring linear blends of images inspired by John F. Kennedy's speech challenging the country to put a man on the moon and the image of Buzz Aldrin standing on the moon as part of the Apollo 11 mission.

The type of linear combination used in this section, $\alpha J + (1 - \alpha)B$, in which the coefficients α and $1 - \alpha$ are both positive and sum to 1 is called a *convex combination*. More broadly, this is a *linear combination*. A linear combination is any sum $xB + yC$ where x and y are scalars. In this case, we enforced that $x + y = 1$. As we'll see in this chapter, convex combinations are also useful in the animation of images.

Before exploring moving images, let's see another example of a convex combination in a still image. Recently, I read a quote of Woody Harrelson, "A grownup is a child with layers on." I'll visualize Harrelson's sentiment

Figure 3.4. A linear blend of images of a cub and adult lion.

using the images of a cub and adult lion in Figures 2.2 (a) and (b). I'll create a blended image column by column. In the blended image, a column of pixels equals the value of the pixels in the same column of the image $\alpha L + (1 - \alpha)C$, where L and C are the images of the adult lion and cub, respectively. We'll take $\alpha = 1$ on the far left and $\alpha = 0$ on the far right. The resulting image can be seen in Figure 3.4.

Note the different ways we've used one mathematical method. In a moment, we will use a convex combination to create even more effects. This connects to an important point about the field and study of mathematics. The boundaries of mathematics often widen through one mathematician building off the work of another. Isaac Newton, a founder of calculus, stated, "If I have seen further than others, it is by standing on the shoulders of giants." One idea can clear the path to another. We see this in this chapter as the idea of a convex combination allows a variety of applications. As you think about the ideas of this section and continue to read, be mindful of your thoughts and ideas. What comes to mind? Can you use these ideas to see further and in other directions?

3.4 Linearly Invisible

Now that we've learned how to do a convex combination, let's learn to cast a hex or spell with it! Our magic wand will be our convex combination.

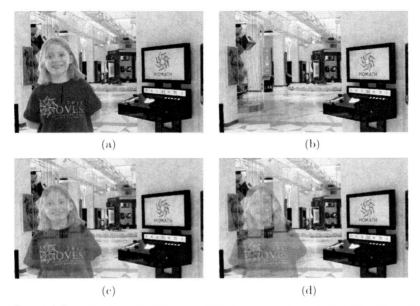

(a) (b)

(c) (d)

Figure 3.5. Mikayla in the Museum of Mathematics (a), made invisible (b), and in stages of becoming invisible (c) and (d). Photographs courtesy of the National Museum of Mathematics.

We will make Mikayla, as seen in Figure 3.5 (a), disappear in the National Museum of Mathematics in New York City.

How do we place such a hex on Mikayla? The trick is letting the beginning matrix X be Figure 3.5 (a) and the final matrix, Y, be Figure 3.5 (b). Then, we visualize the image corresponding to the matrix $N = \alpha Y + (1 - \alpha)X$, for differing values of α. We start with $\alpha = 0$, which corresponds to $N = X$. Then, α is incremented to 0.1, so $N = \alpha Y + (1 - \alpha)X = 0.1Y + 0.9X$. We continue to increment α until $\alpha = 1$ at which point $N = \alpha Y + (1 - \alpha)X = Y$.

What's happening? Consider a pixel that has the same value in Figures 3.5 (a) and (b). Then $x_{ij} = y_{ij}$, where x_{ij} and y_{ij} denote elements in X and Y that appear in row i and column j. Remember $n_{ij} = \alpha y_{ij} + (1 - \alpha)x_{ij} = \alpha x_{ij} + (1 - \alpha)x_{ij} = x_{ij}$. So, the pixel stays the same. For a pixel that is part of Mikayla in 3.5 (a) and the museum in 3.5 (b), then n_{ij} progressively contains more of a contribution from Y as α increases. For example, Figure 3.5 (c) and (d) are formed with $\alpha = 0.25$ and 0.6, respectively.

(a)   (b)

Figure 3.6. Noah on the Davidson College campus and in the Museum of Mathematics. Photographs courtesy of Davidson College and the National Museum of Mathematics.

3.5 Leaving Through a Portal

This same idea can be used for another visual effect. This time, let's start with the image in Figure 3.6 (a) of the Davidson College campus and end with the image in (b) located at the Museum of Mathematics. What visual effect can a convex combination have on Noah who is standing in both pictures? Before reading further, guess what happens if we visualize the matrix N for $N = \alpha Y + (1 - \alpha)X$ where α ranges from 0 to 1.

In Figures 3.6 (a) and (b), Noah is placed in the same location in both images. So, this time the background fades from one image to the other. In Figures 3.7 (a) and (b) we see the image N for $\alpha = 0.25$ and 0.6, respectively.

(a)   (b)

Figure 3.7. Noah in stages of teleporting from the Davidson College campus to the Museum of Mathematics. Photographs courtesy of the National Museum of Mathematics and Davidson College.

With the help of linear algebra, Noah gets to teleport from North Carolina to New York.

In this chapter, we saw simple mathematical operations create interesting visual effects. Matrices are powerful mathematical structures useful in many applications. We will see, in time, that more complicated mathematical operations and methods are also possible. But we saw in this chapter that complexity is not always necessary.

4

Fitting the Norm

Let's talk about distance. If $x = 1$ and $y = 1$ and x changes to 5 and y to 10, then the value of y changed more than the value of x. How do we know this? We measure the distance between 1 and 5 and 1 and 10.

Mathematicians define distance in a variety of ways. Actually, we all do, at least from a mathematical standpoint. Consider two points (x_1, y_1) and (x_2, y_2). Then, as seen in Figure 4.1, one way to measure the distance between the points is $\sqrt{(x_2 - x_1)^2 + (y_2 - y_1)^2}$, as given by the Pythagorean theorem. This is called the *Euclidean distance* between two points (x_1, y_1) and (x_2, y_2).

Euclidean distance isn't as descriptive of your travels if you live in Manhattan and drive a taxi whether it be an automobile today or a horse and buggy over 150 years ago. Consider standing on Broadway between Grand and Broome Streets today or in 1850 as seen in Figure 4.2. Suppose you want to go to a home on the Upper East Side of the city. You could measure how far the points are as a pigeon would fly. This uses Euclidean distance. But, it may be more helpful to know the distance you'd drive on the streets. Take a moment and think about how this distance would be measured. You'd sum the distance traveled in the x direction and the distance in the y direction. This leads to another measure of distance that uses *taxicab geometry*, where the distance between the points (x_1, y_1) and (x_2, y_2) equals $|x_2 - x_1| + |y_2 - y_1|$.

Some applications find the distance between points. Other times, we measure the distance of a point from the origin, which can also be computed with the length of a vector. Vectors are matrices with either one row or one column. For example,

$$\mathbf{v} = \begin{bmatrix} x \\ y \end{bmatrix}$$

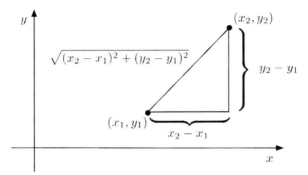

Figure 4.1. Measuring the Euclidean distance between two points.

is a 2D column vector. To keep things compact, we'll often write the transpose of a vector, which turns a column vector into a row vector and vice versa. The transpose of the vector $\mathbf{v}$ is written $\mathbf{v}^T$ and equals $\begin{bmatrix} x & y \end{bmatrix}$. So, $\mathbf{v} = \begin{bmatrix} x & y \end{bmatrix}^T$. We can also take the transpose of a matrix A, which is formed by taking an element in row i and column j of A and placing it in row j and column i of the transpose of A, written A^T.

For a vector $\mathbf{v} = \begin{bmatrix} x & y \end{bmatrix}^T$, the *Euclidean norm* equals $\|\mathbf{v}\|_2 = \sqrt{x^2 + y^2}$. Note that the Euclidean norm of the vector $\mathbf{v}$ is exactly the same as the Euclidean distance between the two points $(0, 0)$ and (x, y).

To find the distance between the points (x_1, y_1) and (x_2, y_2), we can also use a vector norm. Let $\mathbf{u}_1 = \begin{bmatrix} x_1 & y_1 \end{bmatrix}^T$ and $\mathbf{u}_2 = \begin{bmatrix} x_2 & y_2 \end{bmatrix}^T$. Then the distance between the points is analogous to finding the Euclidean norm of

(a)   (b)

Figure 4.2. The image in (a) is the view of New York City looking north on Broadway between Grand and Broome streets in the 1850s. In (b), we see an aerial view of the city at that time.

the vector $\|\mathbf{u}\|_2 = \|\mathbf{u}_1 - \mathbf{u}_2\|$. There is also the *taxicab norm*, which is computed as $\|\mathbf{v}\|_1 = \|\begin{bmatrix} x & y \end{bmatrix}^T\|_1 = |x| + |y|$.

Suppose we are interested in the points $(-1, 5)$ and $(2, 4)$. The distance between them is the length of the vector $\mathbf{u} = \begin{bmatrix} -1 & 5 \end{bmatrix}^T - \begin{bmatrix} 2 & 4 \end{bmatrix}^T = \begin{bmatrix} -3 & 1 \end{bmatrix}^T$, under the vector norm you choose. Using the Euclidean norm, the length of the vector is $\sqrt{(-3)^2 + (1)^2} = \sqrt{10}$, which again correlates to the Euclidean distance between the points. Using the taxicab norm, the length of the vector $\mathbf{u}$ equals $\|\mathbf{u}\|_1 = |-3| + |1| = 4$.

Sometimes, we don't want or need to specify which norm we'll use. In such cases, we can write a general norm by writing $\| \cdot \|$, which is a function that maps vectors in n-dimensional space to the set of real numbers. Regardless of what norm you choose, the norm will have specific properties. In particular, for any two n-dimensional vectors $\mathbf{u}$ and $\mathbf{v}$ and real number a,

1. $\|a\mathbf{v}\| = |a|\|\mathbf{v}\|$.
2. $\|\mathbf{u} + \mathbf{v}\| \leq \|\mathbf{u}\| + \|\mathbf{v}\|$, which is called the triangle inequality.
3. If $\|\mathbf{v}\| = 0$, then $\mathbf{v}$ equals the zero vector.

This allows mathematics to be developed without being specific about a norm. In some cases, you want and need to be specific. Sometimes, you can work generally and later decide if you are using the Euclidean norm, taxicab norm, or another norm of your choosing.

4.1 Recommended Movie

The Euclidean norm can be extended into more than two dimensions. In fact, let's see how to use vector norms in the fourth dimension to find people with similar tastes in movies. Let's look at two of my friends, Oscar and Emmy, and compare their preferences for a few movies to my own. To keep things simple, we'll look at films that were nominated for a 2014 Academy Award for best picture: *American Hustle*, *Gravity*, *Her*, and *Philomena*. Each person rates the films between -5 and 5. A rating of 5 correlates to wanting to see the film (for the first time or again) and -5 means definitely not wanting to see it again. To keep things general, I'll let a random number generator create ratings for all of us. The ratings appear in Table 4.1.

Now, we take the ratings and create preference vectors. For instance, my preferences become the vector $\mathbf{t} = \begin{bmatrix} -3 & 3 & 5 & -4 \end{bmatrix}^T$, Oscar's is $\mathbf{o} =$

Table 4.1. *Movie Ratings for Oscar,*
Emmy, and Me.

	Oscar	Emmy	Tim
American Hustle	0	5	−3
Gravity	5	2	3
Her	−3	5	5
Philomena	4	5	−4

$[\,0 \ 5 \ -3 \ 4\,]^T$ and Emmy's $e = [\,5 \ 2 \ 5 \ 5\,]^T$. To find the preference similarity between Oscar and me, we compute

$$\|\mathbf{o} - \mathbf{t}\|_2 = \sqrt{((-3) - 0)^2 + (3 - 5)^2 + (5 - (-3))^2 + ((-4) - 4)^2}$$

$$= 11.8743.$$

Similarly, the distance between Emmy's and my movie preferences is 12.0830. So, under this norm, my taste is more like Oscar's than Emmy's. The distance between Emmy's and Oscar's movie vectors is 9.9499, so they may enjoy a movie together more, at least if their experience aligns with Euclidean distance.

We just built our vectors from the columns of Table 4.1. Why stop there? Let's create our vectors from the rows of this table, which builds movie vectors. For example, the *American Hustle* vector is $\begin{bmatrix} 0 & 5 & -3 \end{bmatrix}$. With user vectors, we found similar users. With movie vectors, we can find similar movies, at least according to users' ratings. For example, the Euclidean distance between the *American Hustle* and *Philomena* vectors is $4.1231 = \sqrt{(0 - 4)^2 + (5 - 5)^2 + (-3 - (-4))^2}$. If you find the Euclidean distance between every pair of vectors, you'd find these to be the closest.

But, how helpful is this? Each person's ratings were created randomly. What if we take real ratings? Could we mine useful information? Rather than random ratings, let's look at 100,000 ratings of over 900 people with the MovieLens dataset collected by the GroupLens Research Project at the University of Minnesota. This dataset consists of 100,000 ratings (where a rating is between 1 and 5) from 943 users on 1682 movies where each user rated at least 20 movies. The data was collected from September 1997 through April 1998.

Let's create a movie vector where the elements are the ratings of all 943 people. A zero indicates that a movie was not rated. We will find the

Figure 4.3. Examples of handwritten threes.

Euclidean distance between the movies and find those most similar, at least under this measure of distance. First, consider the 1977 film *Star Wars*. The top three most similar films are *Return of the Jedi* (1983), *Raiders of the Lost Ark* (1981), and *The Empire Strikes Back* (1980). How about *Snow White and the Seven Dwarfs* from 1937? It's top three are *Pinocchio* (1940), *Cinderella* (1950), and *Dumbo* (1941).

Later, we will discuss another way to measure distance and also see, toward the end of the book, how to use clustering algorithms to create mathematical genres from such data. This idea could be applied to datasets that contain descriptive statistics on athletes or ratings data on cars, restaurants, or recipes. What comes to mind for you?

4.2 Handwriting at a Distance

In 2012, the United States Postal Service delivered over 160 billion pieces of mail to more than 152 million homes, businesses, and post office boxes in the U.S. To aid in this huge task, the Postal Service, in 1983, began researching how to have computers read addresses. By 1997, the first prototype program correctly identified addresses but only with a success rate of 10 percent for the envelopes it read. At first glance, this may seem poor but it was, in fact, considered a success, given the difficulty of this type of problem. Think about how hard it can be for a human to read some people's handwriting!

Let's see how linear algebra can aid in digit recognition. We will use the MNIST database of 60,000 handwritten digits. Each digit in the database is a 28 × 28 grayscale image. The pictures are oriented so the center of mass of its pixels is at the center of the picture. Figure 4.3 depicts four examples of the number 3. Can we create an algorithm to automatically recognize that each is the digit 3?

We'll create our algorithm with two databases. The first is used to train the algorithm. That dataset contains 60,000 handwritten digits. The second dataset contains new digits, not part of that original 60,000, that we use to see if our algorithm can accurately classify previously unseen digits. We'll

Figure 4.4. The numbers 0 through 9 created through averaging pixel values of handwritten digits.

call these new digits *test digits*. We'll store each image of a handwritten digit as a 784 × 1 vector. So, test digits become *test vectors*. There are a number of ways, some quite advanced, to use the training data to create an identification algorithm. We'll try a very simple one that utilizes vector norms.

The training set contains 5,923 images of the number 0. Rather than compare our test digit to each image, we will create an average handwritten 0 from the training set. The algorithm is fairly simple. Let a pixel in our "average 0" equal the average (or mean) of the pixel in the same location for all zeros in the training set. For example, for the pixel in the upper left hand corner of the "average 0," take the average of the pixel in the upper left hand corner of all zeros in the training set. What does such an average digit look like? You can see all ten, computed for all the numbers from 0 to 9 in Figure 4.4.

We are now ready to classify a test digit. We'll denote the associated test vector by **t**. We'll also denote by $\mathbf{m}_i$ as the vector associated with the average number i, again formed from the pixels of all digits classified as the number i in the training set. To classify our test digit, compute

$$\|\mathbf{m}_i - \mathbf{t}\|_2$$

for $0 \leq i \leq 9$. This computes the distance between the test vector and the vector for each average digit computed above. Whatever value i produces the smallest distance (value of $\|\mathbf{m}_i - \mathbf{t}\|_2$) is our classification for the test digit. For instance, if the smallest norm comes from the distance between the vector for the average 4 and the test digit, then we classify the test digit as a 4.

Figure 4.5. A handwritten 3 that is not correctly identified by our algorithm.

This method correctly classifies all 3s in Figure 4.3. Much more complicated algorithms for digit recognition exist since it can be difficult to fully recognize handwritten digits. Sometimes, we write sloppily and write numerals that are hard to read. For example, the digit in Figure 4.5 is intended to be a 3, but our algorithm classifies it as a 5. See why? The elements of the digit 5 that are within it. While not perfect, this is a simple and accessible algorithm for classifying digits. Do any ideas come to mind as to how you might alter it?

We are now ready to return to arithmetic and learn to multiply, which will give us another way to measure similarity.

5
Go Forth and Multiply

Matrix addition is intuitive. Take two matrices, A and B, that have the same size. If $C = A + B$, then any element in C equals the sum of the elements in the same position in A and B. Scalar multiplication was especially easy. The product $2A$ is obtained by multiplying every element of A by 2. What about multiplying two matrices together? How is this accomplished? Matrix addition allowed one to disappear and scalar multiplication could darken an image. What types of things can matrix multiplication do?

In the first section of this chapter, we will learn to multiply a column and a row vector and then, in the following section, see an application. Then, we will learn to multiply a matrix by a column vector and learn applications in the following sections. So, let's start multiplying in three dimensions and soon go to 4D and beyond.

5.1 Scaly by Product

A foundation of multiplying matrices is understanding how to multiply a row vector by a column vector. To be multiplied, the two vectors must have the same number of elements. So, a 1×4 row vector cannot be multiplied by a 3×1 column vector. For an example, let's find the product

$$\begin{bmatrix} 4 & 1 & 3 \end{bmatrix} \begin{bmatrix} 5 \\ 6 \\ 8 \end{bmatrix},$$

which results in a single number, a scalar.

To find the product, multiply the first elements in each vector. Then add to this the product of the second elements of each vector. We continue until we've summed the product of element i of each vector for every element.

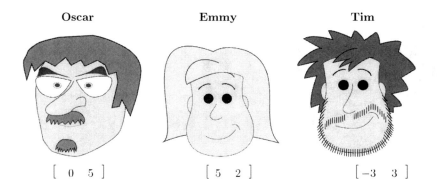

Figure 5.1. Oscar, Emmy, and me and our associated movie ratings as vectors.

Trying this on our product, we find the result to be $(4)(5) + (1)(6) + (3)(8) = 20 + 6 + 24 = 50$.

We can perform a similar computation on two column vectors. If we have $\mathbf{u} = \begin{bmatrix} 4 & 1 & 3 \end{bmatrix}^T$ and $\mathbf{v} = \begin{bmatrix} 5 & 6 & 8 \end{bmatrix}^T$ then we can find the *scalar product* or *dot product* of $\mathbf{u}$ and $\mathbf{v}$, written $\mathbf{u} \cdot \mathbf{v}$, by computing $\mathbf{u}^T \mathbf{v}$. It can be helpful to know and use properties of the dot product,

$$\mathbf{u} \cdot \mathbf{v} = \mathbf{v} \cdot \mathbf{u},$$
$$\mathbf{u} \cdot (\mathbf{v} + \mathbf{w}) = (\mathbf{u} \cdot \mathbf{v}) + (\mathbf{u} + \mathbf{w}),$$
$$k(\mathbf{u} \cdot \mathbf{v}) = (k\mathbf{u}) \cdot \mathbf{v} = \mathbf{u} \cdot (k\mathbf{v}), \text{ and}$$
$$\mathbf{u} \cdot \mathbf{u} = (\|\mathbf{u}\|_2)^2,$$

for vectors $\mathbf{u}$ and $\mathbf{v}$ of the same length, and real number k.

5.2 Computing Similar Taste

The dot product can be used to identify people with similar tastes, in a manner similar to our use of vector norms in Section 4.1. This is a different measure that perceives closeness from a different (mathematical) perspective. As before, we need to vectorize each person's taste. To do this, we'll again use a questionnaire with numeric answers.

Let's return to the movie preferences selected by Oscar, Emmy, and me but now only for the films *American Hustle* and *Gravity*. Keeping only two ratings will enable us to graph the points. Again, we rate movies between -5 and 5 with 5 correlating to really wanting to see the film and -5 being definitely not seeing it.

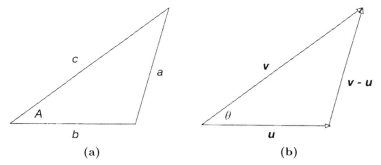

Figure 5.2. The law of cosines helps derive a formula for the angle between vectors.

Oscar rates *American Hustle* with a 0 and *Gravity* with a 5. Emmy rates the films 5 and 2, and I rate them -3 and 3. Each person's ratings again becomes a vector. So, Oscar's taste vector is $\begin{bmatrix} 0 & 5 \end{bmatrix}$. Emmy's vector is $\begin{bmatrix} 5 & 2 \end{bmatrix}$, and mine is $\begin{bmatrix} -3 & 3 \end{bmatrix}$.

If we used our earlier method from Section 4.1, we would find the value of $\|\mathbf{u} - \mathbf{v}\|$, where $\mathbf{u}$ and $\mathbf{v}$ are taste vectors. The two vectors that result in the smallest value are identified as most similar. This time, we'll measure similarity as the angle between the taste vectors.

In 2D, the formula can be derived from the law of cosines which for Figure 5.2 (a) states

$$a^2 = b^2 + c^2 - 2bc \cos A.$$

This can be rearranged as

$$\cos A = \frac{-a^2 + b^2 + c^2}{2bc}.$$

Now, let's view the sides of the triangle as vectors, which we see in Figure 5.2 (b). Note how we are viewing the addition and subtraction of vectors graphically in this picture. The law of cosines states

$$\|\mathbf{v} - \mathbf{u}\|_2^2 = \|\mathbf{u}\|_2^2 + \|\mathbf{v}\|_2^2 - 2\|\mathbf{u}\|_2\|\mathbf{v}\|_2 \cos \theta.$$

We can write

$$
\begin{aligned}
\|\mathbf{v} - \mathbf{u}\|_2^2 &= (\mathbf{v} - \mathbf{u}) \cdot (\mathbf{v} - \mathbf{u}) \\
&= \mathbf{v} \cdot \mathbf{v} - 2\mathbf{v} \cdot \mathbf{u} + \mathbf{u} \cdot \mathbf{u} \\
&= \|\mathbf{v}\|_2^2 - 2\mathbf{v} \cdot \mathbf{u} + \|\mathbf{u}\|_2^2,
\end{aligned}
$$

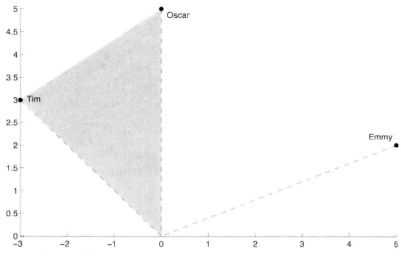

Figure 5.3. Oscar's, Emmy's, and my movie ratings graphed in 2D.

since $\mathbf{u} \cdot \mathbf{u} = \|\mathbf{u}\|^2$. Returning to the law of cosines, we get

$$\|\mathbf{v}\|_2^2 - 2\mathbf{v} \cdot \mathbf{u} + \|\mathbf{u}\|_2^2 = \|\mathbf{u}\|^2 + \|\mathbf{v}\|_2^2 - 2\|\mathbf{u}\|_2\|\mathbf{v}\|_2 \cos\theta.$$

Cancellation and a bit of rearrangement gives us the formula

$$\cos\theta = \frac{\mathbf{u} \cdot \mathbf{v}}{\|\mathbf{u}\|_2\|\mathbf{v}\|_2},$$

which is called *cosine similarity*. The numerator is the scalar product of the vectors and the denominator is the product of their lengths.

Using this formula, the angle between Oscar's and my taste vectors is 45 degrees and Emmy's and my vectors is 113 degrees. Euclidean distance measured length with a ruler. Cosine similarity measures length with a protractor and is called angular distance. The smaller the angle, the closer the vectors are in terms of angular distance. So, Oscar and I are closer to each other, which matches us up, mathematically, as movie buddies!

But wait, Oscar's and my movie tastes are closest under the Euclidean norm, too. Is there any difference between the measures? Let's look at things graphically to see how they can lead to different measures of similarity for data. The graph in Figure 5.3 has ratings for *American Hustle* as the x-axis and *Gravity* as the y-axis. Oscar's, Emmy's, and my preferences appear on the graph.

Euclidean distance can be seen as finding the length of the line segment connecting a pair of points. Cosine similarity measures the angle between the lines connecting the points to the origin. In Figure 5.3, the dotted lines depict the vectors for each movie preference. The angle between Oscar's and my vectors, which is shaded in the graph, is smaller than the angle between Oscar's and Emmy's preference vectors or Emmy's and mine. We also see that Euclidean distance measures Oscar's and my preferences to be closest, too.

The two distance measures will not always be the same. Suppose our ratings were $\begin{bmatrix} 1 & 1 \end{bmatrix}$, $\begin{bmatrix} 5 & 4 \end{bmatrix}$, and $\begin{bmatrix} 5 & 5 \end{bmatrix}$. Euclidean distance measures $\begin{bmatrix} 5 & 4 \end{bmatrix}$ and $\begin{bmatrix} 5 & 5 \end{bmatrix}$ as the closest. Cosine similarity, on the other hand, measures the angle between $\begin{bmatrix} 1 & 1 \end{bmatrix}$ and $\begin{bmatrix} 5 & 5 \end{bmatrix}$ as zero. At first glance, this may seem to indicate that cosine similarity is useless in this setting. Does someone who rates two movies as 5 and 5 really want to see a movie with someone who rated those same movies as 1 and 1?

To help answer this question, let's return to the MovieLens datasets from Section 4.1, which consists of 100,000 ratings from 943 users on 1,682 movies. Our previous example used a vector in 2D because we had two ratings. Now, we create a preference vector with the rating of a movie by every user. So a movie preference vector exists in 943 dimensions. A main question becomes whether to measure similarity using Euclidean distance between vectors, $\|\mathbf{u} - \mathbf{v}\|_2$, or the angle between the vectors with cosine similarity.

In Section 4.1, we found similar movies using Euclidean distance. Now, let's use cosine similarity. For the 1977 film *Star Wars*, the three closest films, as measured with cosine similarity, are *Return of the Jedi* (1983), *Raiders of the Lost Ark* (1981), and *The Empire Strikes Back* (1980). These movies were the closest under Euclidean distance, too. For *Snow White and the Seven Dwarfs*, the closest film is *Beauty and the Beast* from 1991. Then comes *Cinderella* (1950) and *Pinocchio* (1940). Under Euclidean distance, the closest movies were *Pinocchio* (1940), *Cinderella* (1950), and then *Dumbo* (1941).

To underscore the difference in these measures, let's try one more film, *The Princess Bride* from 1987. Cosine similarity yields *Raiders of the Lost Ark* as the most similar. Euclidean distance gives *Monty Python and the Holy Grail* from 1974. Such differences underscore that they glean different information and model similarity differently. Which is better would depend, in part, on the intended use of the information.

5.3 Scaling to Higher Dimensions

We've been multiplying vectors, which are matrices with only one row or one column. Now, let's learn to multiply two matrices without such a restriction. We need to know that AB does not necessarily equal BA. In fact, while it might be possible to calculate AB, BA might not be computable. We know when matrix multiplication is possible by looking at the dimensions of the matrices in the product. If A is $m \times n$ and B is $p \times q$, then n must equal p in order to perform AB; when $n = p$, the resulting matrix is $m \times q$. The number of columns of A must equal the number of rows of B to perform AB. For example, if

$$A = \begin{bmatrix} -1 & 2 & 0 \\ 3 & -2 & 3 \\ 1 & 0 & 4 \end{bmatrix} \quad \text{and} \quad B = \begin{bmatrix} 2 & 1 \\ -1 & 3 \\ 0 & 4 \end{bmatrix},$$

then AB is computable since A has three columns and B has three rows. The product will be a matrix with three rows and two columns. The element in row i and column j of C, where $C = AB$, is found by computing the scalar product of row i of A with the jth column of B. So, the element in the first row and first column of C equals

$$\begin{bmatrix} -1 & 2 & 0 \end{bmatrix} \begin{bmatrix} 2 \\ -1 \\ 0 \end{bmatrix} = -2 - 2 + 0 = -4.$$

We do this type of multiplication to calculate every element in C and find

$$C = \begin{bmatrix} -4 & 5 \\ 8 & 9 \\ 2 & 17 \end{bmatrix}.$$

Computing BA is not possible since B has two columns and A has three rows.

5.4 Escher in the Matrix

Armed with the ability to multiply matrices together, let's take a spin at tiling the plane. Consider the tile in Figure 5.4 constructed with ten line segments. We'll call this the model bird. The ten endpoints of the line segments are $(0.3036, 0.1960)$, $(0.6168, 0.2977)$, $(0.7128, 0.4169)$, $(0.7120, 0.1960)$,

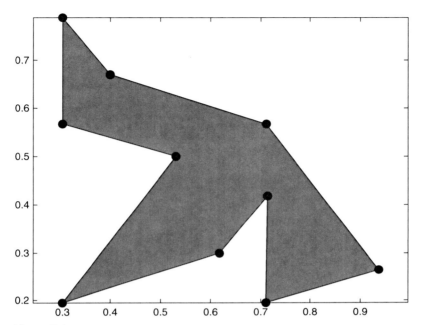

Figure 5.4. A polygon that forms a bird that will tile the plane. Design courtesy of Alain Nicolas.

(0.9377, 0.2620), (0.7120, 0.5680), (0.3989, 0.6697), (0.3028, 0.7889), (0.3036, 0.5680), and (0.5293, 0.5020).

Let's take four tiles of this model bird and fit them together. The first tile is created by rotating the model bird 180 degrees about the point (0.7120, 0.4320). The second tile is formed by flipping (or reflecting) the model bird about the horizontal line $y = 0.6180$ and then translating the image by 0.4084 units along the x-axis. The third tile again reflects the model bird but this time about the vertical line $x = 0.5078$ and translates the image by -0.3720 units along the y-axis. Finally, we reflect once more, this time about the vertical line $x = 0.5078$ and then translate along the y-axis by -0.3720 units. Graphing all four tiles produces Figure 5.5 (a).

If we repeat this pattern of four birds but translate the entire pattern by 0.7441 along the y-axis, we produce a column of the tilings. If we translate by -0.8168 along the x-axis, then we create a row of tiling. A series of horizontal and vertical translations tile the plane as seen in Figure 5.5 (b).

How do we use linear algebra to this? Linear algebra actually makes this easy. To rotate a point (x, y) about the origin, we take the vector $\begin{bmatrix} x & y \end{bmatrix}^T$

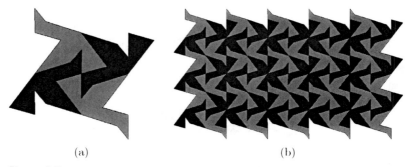

(a)   (b)

Figure 5.5. A pattern of four birds from Figure 5.4 (a) and the pattern repeated to tile the plane (b).

and calculate the vector associated with the rotated point as the product

$$\begin{bmatrix} \cos\theta & -\sin\theta \\ \sin\theta & \cos\theta \end{bmatrix} \begin{bmatrix} x \\ y \end{bmatrix}.$$

To reflect a point about the y-axis, we calculate

$$\begin{bmatrix} -1 & 0 \\ 0 & 1 \end{bmatrix} \begin{bmatrix} x \\ y \end{bmatrix}.$$

Reflecting about the x-axis, we calculate

$$\begin{bmatrix} 1 & 0 \\ 0 & -1 \end{bmatrix} \begin{bmatrix} x \\ y \end{bmatrix}.$$

Finally, translating by h along the x-axis and k along the y-axis is found by computing

$$\begin{bmatrix} h \\ k \end{bmatrix} + \begin{bmatrix} x \\ y \end{bmatrix}.$$

But, wait: translation just introduced addition into the process. This may seem simple, as matrix addition was easy enough. But, performing, for instance, a rotation and translation would involve addition *and* multiplication. Things are simplified if we can compute all our transformations with matrix multiplication as everything could then be consolidated into one matrix multiplication, even if we are rotating, translating, and reflecting.

This goal is achieved if we work in homogeneous coordinates, which are used in projective geometry. The projective plane can be thought of as the Euclidean plane with additional points. Given a point (x, y) on the Euclidean plane, then for a non-zero real number z, (xz, yz, z) are the homogeneous coordinates of the point. For example, the Cartesian point $(1, 3)$ can be represented in homogeneous coordinates as $(1, 3, 1)$ or $(2, 6, 2)$. The original Cartesian coordinates can be recovered by dividing the first two positions by the third. Any single point in Cartesian coordinates can be represented by infinitely many homogeneous coordinates.

Let's see how easy it is to take the model bird and perform each of the maneuvers. If we want to translate a point $(x, y, 1)$ by h along the x-axis and k along the y-axis, we multiply the point by the matrix

$$\begin{bmatrix} 1 & 0 & h \\ 0 & 1 & k \\ 0 & 0 & 1 \end{bmatrix}$$

since

$$\begin{bmatrix} 1 & 0 & h \\ 0 & 1 & k \\ 0 & 0 & 1 \end{bmatrix} \begin{bmatrix} x \\ y \\ 1 \end{bmatrix} = \begin{bmatrix} x + h \\ y + k \\ 1 \end{bmatrix}.$$

A rotation by θ degrees about the origin corresponds to multiplying the point by

$$\begin{bmatrix} \cos\theta & -\sin\theta & 0 \\ \sin\theta & \cos\theta & 0 \\ 0 & 0 & 1 \end{bmatrix}.$$

As seen in Figure 5.6 (a), consider rotating a triangle by θ degrees about a point (a, b). Rotating about a point can be viewed as translating the point (a, b) to the origin, rotating and then translating from the origin back to (a, b). This equates to multiplying by

$$\begin{bmatrix} 1 & 0 & a \\ 0 & 1 & b \\ 0 & 0 & 1 \end{bmatrix} \begin{bmatrix} \cos\theta & -\sin\theta & 0 \\ \sin\theta & \cos\theta & 0 \\ 0 & 0 & 1 \end{bmatrix} \begin{bmatrix} 1 & 0 & -a \\ 0 & 1 & -b \\ 0 & 0 & 1 \end{bmatrix}$$

$$= \begin{bmatrix} \cos\theta & -\sin\theta & -a\cos\theta + b\sin\theta + a \\ \sin\theta & \cos\theta & -a\sin\theta - b\cos\theta + b \\ 0 & 0 & 1 \end{bmatrix}.$$

As an example, to rotate a point 180 degrees about the point $(0.7120, 0.4320)$, we multiply the homogeneous coordinates of the point

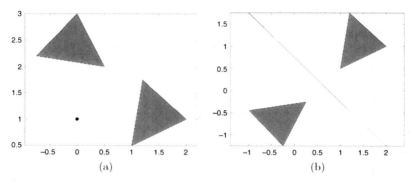

Figure 5.6. Rotating a triangle by 90 degrees about a point (a) and reflecting a triangle about a line (b).

by the matrix

$$\begin{bmatrix} -1 & 0 & 1.4240 \\ 0 & 1 & 0.8640 \\ 0 & 0 & 1 \end{bmatrix}.$$

To reflect a point about the x- or y-axis, we multiply with the matrices

$$\begin{bmatrix} 1 & 0 & 0 \\ 0 & -1 & 0 \\ 0 & 0 & 1 \end{bmatrix} \quad \text{or} \quad \begin{bmatrix} -1 & 0 & 0 \\ 0 & 1 & 0 \\ 0 & 0 & 1 \end{bmatrix},$$

respectively.

More generally, to reflect about the line $ax + by + c = 0$, we multiply by the matrix

$$\begin{bmatrix} b^2 - a^2 & -2ab & -2ac \\ -2ab & a^2 - b^2 & -2bc \\ 0 & 0 & a^2 + b^2 \end{bmatrix}.$$

A reflection about the line $2x + 2y - 1.5 = 0$ is seen in Figure 5.6 (b), which corresponds to multiplying with the matrix

$$\begin{bmatrix} 0 & -8 & 6 \\ -8 & 0 & 6 \\ 0 & 0 & 8 \end{bmatrix}.$$

This gives us all the linear algebra we need to rotate in 2D and tile the plane with the model bird. When tiling, one is left with such decorating decisions as a choice of color and the amount of surface area to cover.

Having tiled in 2D, let's step up a dimension and rotate in 3D.

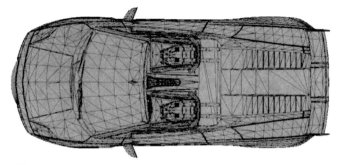

Figure 5.7. A model of a Lamborghini created for 3D printing available on thingi-verse.com. Model by jarred1997, (www.thingiverse.com/thing:125339), CC BY 3.0 (creativecommons.org/licenses/by/3.0/).

5.5 Lamborghini Spinout

From movie animation to 3D printing, wireframe models play an instrumental role in digitally creating an object in a computer. Such models are also used in simulation where computational fluid dynamics allows one to study the aerodynamics of an object, such as a car. Keeping with the auto theme, let's see a model of a Lamborghini Gallardo visualized in Figure 5.7.

Why use such a wireframe model? One reason is the ease of movement. Let's explore this idea using the 3D model of the Lamborghini. The model contains 312,411 vertices. To produce it, we need to know the 3D coordinates of every point and which vertices are connected with lines.

Let's store the coordinates of each vertex in a row of the matrix V. So, V would have three rows and 312,411 columns for the Lamborghini in Figure 5.7. The image can be rotated by θ degrees about the z-axis by performing the matrix multiplication $R_z(V)$ where

$$R_z = \begin{bmatrix} \cos\theta & -\sin\theta & 0 \\ \sin\theta & \cos\theta & 0 \\ 0 & 0 & 1 \end{bmatrix}.$$

Rotating the point $(1, 0, 0)$ by 90 degrees about the z-axis leads to

$$\begin{bmatrix} 0 & -1 & 0 \\ 1 & 0 & 0 \\ 0 & 0 & 1 \end{bmatrix} \begin{bmatrix} 1 \\ 0 \\ 0 \end{bmatrix} = \begin{bmatrix} 0 \\ 1 \\ 0 \end{bmatrix}.$$

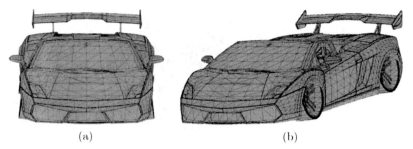

(a)   (b)

Figure 5.8. A model of a Lamborghini created with 312,411 vertices. A rotation of a wireframe involves rotating the vertices in (a) and redrawing the edges between vertices to form the tessellation (b). Model by jarred1997, (www.thingiverse.com/thing:125339), CC BY 3.0 (creativecommons.org/licenses/by/3.0/).

Rotating by θ degrees about the x-axis is performed by multiplying by

$$R_x = \begin{bmatrix} 1 & 0 & 0 \\ 0 & \cos\theta & -\sin\theta \\ 0 & \sin\theta & \cos\theta \end{bmatrix}.$$

For a wireframe model, each point is repositioned by multiplying the vector associated with the point by the rotation matrix. Then lines are drawn between connected points. In Figure 5.8 (a), the model of the car in Figure 5.7 is rotated by $\theta = 90$ degrees around the z-axis and then about 82 degrees about the x-axis.

We'll rotate the model about the y-axis so we see the side of the car. This is performed with a rotation about the y-axis, which is computed by multiplying the matrix

$$R_y = \begin{bmatrix} \cos\theta & 0 & \sin\theta \\ 0 & 1 & 0 \\ -\sin\theta & 0 & \cos\theta \end{bmatrix}.$$

The model pictured in Figure 5.8 (b) rotated the model seen in Figure 5.8 (a) by -25 degrees about the y-axis.

The matrices V and R_y are $3 \times 312,411$ and 3×3, respectively. One rotation of the image requires $(312,411)(3)(3) = 2,811,699$ multiplications. Modern computers can perform such a computation very quickly. This is important as creating the images in Figure 5.8 required the computer to also calculate what is and isn't viewable since the polygons are opaque.

5.6 Line Detector

This chapter started with the topic of the dot product and progressed toward matrix multiplication as a tool for 3D models. Now, let's conclude the chapter with an application of the dot product in graphics.

First, let's show that if $y = f(x)$ is linear, then

$$f(x) = \frac{f(x+h) + f(x-h)}{2}.$$

Since $y = f(x)$ is linear, we can express it as $y = mx + b$. So, we have

$$\frac{f(x+h) + f(x-h)}{2} = \frac{m(x+h) + b + m(x-h) + b}{2}$$

$$= \frac{2mx + 2b}{2} = mx + b = f(x).$$

Moving $f(x)$ to the other side of the equation gives

$$0 = \frac{f(x+h) - 2f(x) + f(x-h)}{2}.$$

Let's define a new function g as

$$g(x) = \frac{f(x+h) - 2f(x) + f(x-h)}{h^2}.$$

From what we just showed, if f is a line then $g(x) = 0$. So, if $f(x) = 2x + 1$, then again $g(x) = 0$.

This will help us detect edges in a picture. How? First, take $h = 1$. Then

$$g(x) = f(x+1) - 2f(x) + f(x-1).$$

Suppose we have a gray line of three pixels formed with grayscale values 85, 85, and 85. This would create the pixel image seen in Figure 5.9 (a). Let's apply our formula to the middle pixel, letting $f(x)$ equal the value of the middle pixel and $f(x+1)$ and $f(x-1)$ equal the values of the right and left pixels, respectively. Then, $g(x) = 85 - 2(85) + 85 = 0$, which is black.

Suppose further that we have the values $35, 85$, and 135, which together create the three-pixel image seen in Figure 5.9 (b). The pixel values were generated using the formula $50(x-1) + 35$, where $x = 0, 1, 2$, respectively. Thus, their values increase linearly. Applying our formula in the same way to find $g(x)$ where $f(x)$ is the value of the middle pixel, $g(x) = 35 + 2(85) - 135 = 70$.

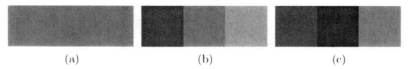

(a)   (b)   (c)

Figure 5.9. Three three–pixel images.

Finally, let's take three nonlinear pixel values 50, 30, 103 that we see visualized in Figure 5.9 (c). Then, $g(x) = 50 - 2(30) + 103 = 93$. Keep in mind, we could get negative values for $g(x)$. If this happens, we will then replace $g(x)$ by its absolute value.

We applied the formula to one pixel. Now, we will apply it to every pixel in an n by m matrix of pixel values. We will call the pixel in row i and column j, pixel (i, j). We can apply our formula to detect changes in color in the horizontal direction by computing

$$P = (\text{value of pixel } (i, j + 1)) - 2(\text{value of pixel } (i, j))$$
$$+(\text{value of pixel } (i, j - 1)).$$

For instance, suppose the pixel at row i and column j has a grayscale value of 25. The pixel to the left of it has a value of 40 and the pixel to the right it has a value of 20. Then $P = 40 - 2(25) + 20 = 10$. We will then replace the pixel in row i and column j by the value of P. Doing this for every pixel in the image forms a new image. If a pixel in the original image had the same value as the pixel to the right and left of it, then the new value will be 0. Similarly, if the color at a pixel changes linearly in the horizontal direction, then it will be colored black in the new image. Again, we take the absolute value of this computation, which is not a linear process but is the only nonlinear step. There is the question of what values to set for the pixels on the top and bottom rows. In computer graphics, an option, which is the one we'll take, is to set these pixels to black.

Where's the linear algebra? Our computation is, in fact, a dot product. For the example, if we are interested in a pixel with values 40, 25, and 20 then the new image would have the pixel in the same location colored with the value of $\begin{bmatrix} 40 & 25 & 20 \end{bmatrix} \cdot \begin{bmatrix} 1 & -2 & 1 \end{bmatrix}$. More generally, let $\mathbf{u}_{i,j} = \begin{bmatrix} p_{i,j+1} & p_{i,j} & p_{i,j-1} \end{bmatrix}$, where $p_{i,j}$ equals the value of the pixel (i, j) in the image. Then, we replace pixel $p_{i,j}$ in the image with $\mathbf{u}_{i,j} \cdot \begin{bmatrix} 1 & -2 & 1 \end{bmatrix}$.

What does this look like for an image? Applying this technique to the grayscale image of the Mona Lisa in Figure 5.10 (a) produces the image in

(a)   (b)   (c)

Figure 5.10. The grayscale image of the Mona Lisa (a) undergoes horizontal edge detection (b) with the image in (b) inverted to produce (c).

Figure 5.10 (b). An interesting variation is to invert the colors in (b), which results in the image in Figure 5.10 (c).

In this chapter, we have seen applications of multiplication. Multiplying two vectors allows us to compute similarities in taste and create an image edge detector. Multiplying two matrices allowed us to tile the plane and rotate a wireframe. In the next chapter, we work with operations that allow us to solve some of the algebra problems in linear algebra. Multiplication shows us how to find $\mathbf{b}$ by computing $A\mathbf{x}$. The next chapter relates to the mechanics of finding $\mathbf{x}$ if we are given A and $\mathbf{b}$ and we know $A\mathbf{x} = \mathbf{b}$.

6
It's Elementary, My Dear Watson

We've considered modern applications of linear algebra in this book—from computer graphics to preferences in movies. Applying linear algebra isn't new, though. Techniques to solve such problems have been studied for some time. Let's step back to the era when the Great Wall of China was completed, 200 BC. This is the same period to which a Chinese manuscript, the *Jiuzhang Suanshu*, has been dated. The ancient document contains 246 problems intended to illustrate methods of solution for everyday problems in areas such as engineering, surveying, and trade. The eighth chapter details the first known example of matrix methods with a technique known as *fangcheng*, which is what would become known centuries later as Gaussian elimination and is a topic of this chapter. The last chapter showed us how to compute **b** where $A\mathbf{x} = \mathbf{b}$. This chapter shows us how to find **x** when $A\mathbf{x} = \mathbf{b}$ which again was detailed with the *fangcheng* over two millenia ago.

Let's return to a problem we saw at the end of Chapter 1. It's from the eighth chapter of the *Jiuzhang Suanshu*. Momentarily, we'll learn how to solve it using Gaussian elimination.

> There are three classes of grain, of which three bundles of the first class, two of the second, and one of the third make 39 measures. Two of the first, three of the second and, one of the third make 34 measures. And one of the first, two of the second and three of the third make 26 measures. How many measures of grain are contained in one bundle of each class?

This problem becomes

$$\begin{aligned} x &+ 2y &+ 3z &= 26 \\ 3x &+ 2y &+ z &= 39 \\ 2x &+ 3y &+ z &= 34 \end{aligned}$$

which has the solution $x = 9.25$, $y = 4.25$, and $z = 2.75$ for the measures of grain in the three classes. Did you get it?

Let's see how to solve this type of system. Essentially, one eliminates variables by adding or subtracting appropriate multiples of one equation to another. For example, to eliminate the variable x from the second equation, one could subtract 3 times the first equation from the second to obtain

$$(3x + 2y + z) - 3(x + 2y + 3z) = 39 - 3(26), \quad \text{or} \quad -4y - 8z = -39.$$

To eliminate x from the third equation, one could subtract 2 times the first equation from the third and get

$$(2x + 3y + z) - 2(x + 2y + 3z) = 34 - 2(26), \quad \text{or} \quad -y - 5z = -18.$$

Then, we could eliminate y from this last equation by multiplying that last equation by 4 and subtracting it from the previous one:

$$(-4y - 8z) - 4(-y - 5z) = -39 - 4(-18), \quad \text{or} \quad 12z = 33.$$

From this it follows that $z = 2.75$, and substituting this value into the previous equation gives $-y - 5(2.75) = -18$, or $y = 4.25$. Finally, substituting these values into the first equation, we find $3x + 2(4.25) + 2.75 = 39$ or, $x = 9.25$.

In matrix form, we express the system of linear equations as

$$\begin{bmatrix} 1 & 2 & 3 \\ 3 & 2 & 1 \\ 2 & 3 & 1 \end{bmatrix} \begin{bmatrix} x \\ y \\ z \end{bmatrix} = \begin{bmatrix} 26 \\ 39 \\ 34 \end{bmatrix}.$$

Just as before, we can subtract 3 times the first row from the second, which produces the matrix system

$$\begin{bmatrix} 1 & 2 & 3 \\ 0 & -4 & -8 \\ 2 & 3 & 1 \end{bmatrix} \begin{bmatrix} x \\ y \\ z \end{bmatrix} = \begin{bmatrix} 26 \\ -39 \\ 34 \end{bmatrix}.$$

Following the same steps as before, we'd keep adding and subtracting multiplies of rows to other rows until we get the matrix system

$$\begin{bmatrix} 1 & 2 & 3 \\ 0 & -4 & -8 \\ 0 & 0 & 12 \end{bmatrix} \begin{bmatrix} x \\ y \\ z \end{bmatrix} = \begin{bmatrix} 26 \\ -39 \\ 33 \end{bmatrix}.$$

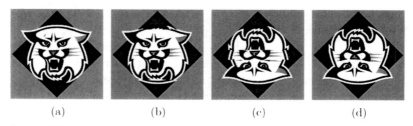

(a) (b) (c) (d)

Figure 6.1. An image of the Davidson College wildcat (a) and the image after swapping 80 (b), 160 (c), and 180 (d) first and last rows.

This has all zeros under the main diagonal of the matrix on the left-hand side. At this point, we can use substitution, just as we did, to find our solution.

This process is called *Gaussian elimination* with *back substitution*.

6.1 Visual Operation

Gaussian elimination uses elementary row operations on a linear system to produce a new system of linear equations with the same solution. As such, one moves from the initial matrix system to one in which back substitution can be performed. The elementary row operations that maintain the same solution are

1. Swap any two rows.
2. Multiply a row by a nonzero number.
3. Change a row by adding to it a multiple of another row.

Let's get a visual sense of these elementary operations.

6.1.1 Flipped Out Over Davidson

First, let's look at swapping any two rows. As we've seen, an image can be viewed as a matrix of colors. We'll use the image of the Davidson College wildcat logo in Figure 6.1 (a) whose dimensions are 360 by 360 pixels. Each entry of the matrix stores a color of the corresponding pixel. This assumes that color is stored as a single value. Often, a color pixel is stored as a triplet indicating the amount of red, green, and blue that combine to produce the color. When stored in this way, we can store the color information of an n by m pixel image with three n by m matrices.

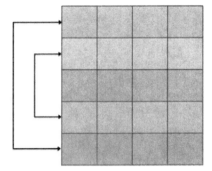

Figure 6.2. Swapping rows in a matrix is an elementary row operation.

Let's flip the Wildcat image by swapping rows. To begin, we'll swap the image's first and last rows. Then, the second and second to last rows are interchanged. We continue to flip the image by swapping rows i and $n - i + 1$ where n is the number of rows in the image. In Figure 6.2, we see how such operations would be performed on an image with five rows and four columns of pixels.

This process of row swapping can be performed with matrix multiplication. Let the matrices R, G, and B contain the red, blue, and green intensities of each pixel in the Wildcat logo's image. Interchanging two rows is common in linear algebra and is useful in Gaussian elimination. To perform this as a linear operation, we take the identity matrix, which is the matrix containing zeros for all off-diagonal elements and ones along the main diagonal. To swap rows i and j in a matrix, we form the matrix P equalling the identity matrix with rows i and j swapped. We see an example below that would swap the first and last rows.

$$P = \begin{bmatrix} 0 & 0 & 0 & \cdots & 1 \\ 0 & 1 & 0 & \cdots & 0 \\ \vdots & \vdots & \vdots & \cdots & \vdots \\ 1 & 0 & 0 & \cdots & 0 \end{bmatrix}.$$

To swap the red intensities of the first and last rows of the image, we take PR. Also computing RG and RB will produce an image with the first and last rows swapped.

You could swap columns rather than rows. You could also swap diagonals of the matrix. But, remember, swapping rows is the process used in Gaussian elimination. As you think about other variations, think about linear

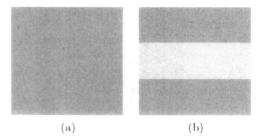

(a) (b)

Figure 6.3. A grayscale image of a box that is 300 by 300 pixels (a) and the image after 100 rows in the middle third are multiplied by 1.5.

operators, like our use of the matrix P in swapping rows, that will perform your visual effect. What variations would you try?

6.1.2 Brightened Eyes

Now, let's look at multiplying a row by a nonzero number. Let's begin with a gray box that is 300 by 300 pixels as seen in Figure 6.3 (a). We'll multiply the rows in the middle third by 1.5 and visualize the image again. We see the new image in Figure 6.3 (b). Remember, larger values correspond to lighter pixel values in grayscale images.

Let's do this again, but now our matrix will be a grayscale image of the Mona Lisa seen in Figure 6.4 (b). The image has 786 rows and 579 columns. Let's multiply every row from row 320 to 420 by 1.5 and then visualize the

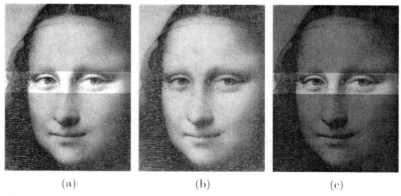

(a) (b) (c)

Figure 6.4. A grayscale image of the Mona Lisa (b) and the image after 100 rows are multiplied by 1.5 (a) and when all but those same 100 rows are multiplied by 0.5 (c).

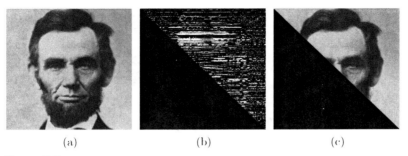

(a)   (b)   (c)

Figure 6.5. If we zero the values below the main diagonal, as done in Gaussian elimination, on an image of Abraham Lincoln, do we get the image in (b) or (c)?

image again. This elementary row operation has brightened Mona Lisa's eyes as seen in Figure 6.4 (a). In contrast, let's leave rows 320 to 420 the same and multiply all other rows by 0.5. This darker image is seen in Figure 6.4 (c).

6.1.3 Substitute Lincoln

Now, let's look at replacing a row by adding it to a multiple of another row. To keep things visual, let's work with the image of Abraham Lincoln in Figure 6.5 (a). Following the steps of Gaussian elimination, we'll place a zero in the first element of every row but the first. Then, we turn to the second column and place a zero in every row except the first two rows. This process is continued for every row until zeros are placed in every element under the main diagonal of the matrix. Now comes a question, would this result in the image in Figures 6.5 (b) or (c)? Note, a black pixel corresponds to a 0 in the matrix. The matrices that produced Figures 6.5 (b) or (c) both have zeros in all elements below their main diagonals. When using Gaussian elimination to solve a linear system, we would back substitute at this stage.

Depending on your intuition with Gaussian elimination, you may have guessed (b) or (c). While the process does produce zeros in all elements below the main diagonal, in most cases the nonzero elements in the row also change. Earlier in this chapter, we had a linear system from the problem posed in the *Jiuzhang Suanshu*,

$$\begin{bmatrix} 1 & 2 & 3 \\ 3 & 2 & 1 \\ 2 & 3 & 1 \end{bmatrix} \begin{bmatrix} x \\ y \\ z \end{bmatrix} = \begin{bmatrix} 26 \\ 39 \\ 34 \end{bmatrix}.$$

After zeroing out all the elements in the first column under the main diagonal, we got the matrix system

$$
\begin{bmatrix} 1 & 2 & 3 \\ 0 & -4 & -8 \\ 0 & -1 & -5 \end{bmatrix} \begin{bmatrix} x \\ y \\ z \end{bmatrix} = \begin{bmatrix} 26 \\ -39 \\ -18 \end{bmatrix}.
$$

The values in the second and last rows changed because they were also affected by the addition of a multiple of the first row.

So, the final process of Gaussian elimination on the image in Figure 6.5 (a) produces the image in (b).

In the last three sections, we've seen three types of row operations needed to solve a linear system. Amazingly, we only need these three operations and no more. At one level, solving a linear system can seem simple. Keep in mind, though, the process of solving a linear system is a common and important procedure in scientific computing. In the next section, we apply linear algebra to cryptography.

6.2 Being Cryptic

Billions of dollars are spent on online purchases resulting in credit card numbers flying through the Internet. For such transactions, you want to be on a secure site. Security comes from the encryption of the data. While essentially gibberish for almost anyone, encrypted data can be decrypted by the receiver. Why can't others decipher it? Broadly speaking, many of the most secure encryption techniques are based on factoring really huge numbers. How big? I'm referring to a number of the size of 1 quattuorvigintillion, which is 10 to the 75th power. Keep in mind that the number of atoms in the observable universe is estimated to be 10 to the 80th power. Such techniques rely heavily on number theory, which as we'll see in this section can overlap with linear algebra.

Let's start with one of simplest encryption methods named the Caesar cipher after Julius Caesar. To begin, letters are enumerated in alphabetical order from 0 to 25 as seen below, which will also be helpful for reference.

a	b	c	d	e	f	g	h	i	j	k	l	m
0	1	2	3	4	5	6	7	8	9	10	11	12

n	o	p	q	r	s	t	u	v	w	x	y	z
13	14	15	16	17	18	19	20	21	22	23	24	25

Let's encrypt the word

LINEAR

which corresponds to the sequence of numbers 11, 8, 13, 4, 0, and 17. The *Caesar cipher*, as it is called, adds 3 to every number. A vector $\mathbf{v}$ is encrypted by computing $\mathbf{v} + 3\mathbf{u}$, where $\mathbf{u}$ is the vector of 1s that is the same size as $\mathbf{v}$. Encrypting $\mathbf{v} = \begin{bmatrix} 11 & 8 & 13 & 4 & 0 & 17 \end{bmatrix}^T$ leads to $\begin{bmatrix} 14 & 11 & 16 & 7 & 3 & 20 \end{bmatrix}^T$. We form the encrypted message by writing the letters associated with the entries of the encrypted vector. For this message we get

OLQHDU

So, the word "linear" becomes "olqhdu," which I, at least, would struggle to pronounce. In cryptography, the receiver must be able to easily decode a message. Suppose you receive the encrypted message

PDWULA

and want to decrypt it. Again, begin by creating the vector associated with these letters, which is $\begin{bmatrix} 15 & 3 & 22 & 20 & 11 & 0 \end{bmatrix}^T$. Decryption on a vector $\mathbf{v}$ is performed by calculating $\mathbf{v} - 3\mathbf{u}$. For our message, the decrypted word vector is $\begin{bmatrix} 12 & 0 & 19 & 17 & 8 & -3 \end{bmatrix}^T$.

Simple enough but we aren't quite done. Our subtraction gave us a negative number. Which letter is this? A similar issue arises if we add 3 to 23 giving 26. When we reach 26, we want to loop around to 0. Returning to the letters, we want a shift of 1 to take us from the letter "z" to the letter "a". In this way, we are working modulo 26, meaning that any number larger than 25 or smaller than 0 becomes the remainder of that number divided by 26. So, -3 corresponds to the same letter as $-3 + 26 = 23$. Now we're ready to find our decrypted message since our decrypted vector, modulo 26, equals $\begin{bmatrix} 12 & 0 & 19 & 17 & 8 & 23 \end{bmatrix}^T$. This corresponds to the letters

MATRIX.

Ready to start passing encrypted notes or sending encrypted text messages? In this example, we shifted by 3. You could just as easily shift by 11, 15, or 22. Just let your receiver know and you are ready to send encrypted messages to a friend.

But, don't get too confident and send sensitive information. This cipher is fairly easy to break, regardless of your shift. If you suspect or know this type of cipher was used, you could try all shifts between 1 and 25 and see if you get a recognizable message. That's especially fast to try on a computer. Often, we encrypt messages so they can't be easily broken. So, let's try a more complicated cipher.

Rather than adding a vector to another vector, we'll multiply a vector by a matrix to get the encrypted vector. This time we'll break our vector into groups of 3 before encoding. For this example, we'll multiply by the matrix

$$\begin{bmatrix} -3 & -3 & -4 \\ 0 & 1 & 1 \\ 4 & 3 & 4 \end{bmatrix}.$$

This will result in the sequence of numbers for the encrypted message.

Let's again encrypt the word

LINEAR

which, as we found, corresponds to the vector $\begin{bmatrix} 11 & 8 & 13 & 4 & 0 & 17 \end{bmatrix}^T$. We break this vector into vectors of length 3 and treat each as a column vector. First, we compute

$$\begin{bmatrix} -3 & -3 & -4 \\ 0 & 1 & 1 \\ 4 & 3 & 4 \end{bmatrix} \begin{bmatrix} 11 \\ 8 \\ 13 \end{bmatrix} = \begin{bmatrix} -109 \\ 21 \\ 120 \end{bmatrix},$$

which modulo 26 is $\begin{bmatrix} 21 & 21 & 16 \end{bmatrix}^T$. You can verify that $\begin{bmatrix} 4 & 0 & 17 \end{bmatrix}^T$ becomes $\begin{bmatrix} 24 & 17 & 6 \end{bmatrix}^T$. Combining the encoded vectors, we form the vector $\begin{bmatrix} 4 & 0 & 17 \\ 24 & 17 & 6 \end{bmatrix}^T$, which corresponds to the encrypted message:

VVQYRG

How do you decode this message? It needs to be a quick and easy process for the receiver. First, of course, we would transfer from the letters to find the first encrypted vector $\begin{bmatrix} 21 & 21 & 16 \end{bmatrix}^T$. To decrypt, we want to find the vector **x** such that

$$\begin{bmatrix} -3 & -3 & -4 \\ 0 & 1 & 1 \\ 4 & 3 & 4 \end{bmatrix} \mathbf{x} = \begin{bmatrix} 21 \\ 21 \\ 16 \end{bmatrix}.$$

Not surprisingly, this is precisely the topic of this chapter. Elementary row operations could be used within Gaussian elimination to solve this linear system and find $\mathbf{x} = \begin{bmatrix} 37 & 216 & -195 \end{bmatrix}^T$. Replacing each of these entries with the remainder when divided by 26 gives us $\begin{bmatrix} 11 & 8 & 13 \end{bmatrix}^T$.

It is also helpful here to use a matrix inverse. Note that

$$\begin{bmatrix} -3 & -3 & -4 \\ 0 & 1 & 1 \\ 4 & 3 & 4 \end{bmatrix} \begin{bmatrix} 1 & 0 & 1 \\ 4 & 4 & 3 \\ -4 & -3 & -3 \end{bmatrix} = \begin{bmatrix} 1 & 0 & 1 \\ 4 & 4 & 3 \\ -4 & -3 & -3 \end{bmatrix} \begin{bmatrix} -3 & -3 & -4 \\ 0 & 1 & 1 \\ 4 & 3 & 4 \end{bmatrix}$$

$$= \begin{bmatrix} 1 & 0 & 0 \\ 0 & 1 & 0 \\ 0 & 0 & 1 \end{bmatrix}.$$

So

$$\begin{bmatrix} 1 & 0 & 1 \\ 4 & 4 & 3 \\ -4 & -3 & -3 \end{bmatrix}$$

is said to be the inverse of

$$\begin{bmatrix} -3 & -3 & -4 \\ 0 & 1 & 1 \\ 4 & 3 & 4 \end{bmatrix}.$$

This helps because

$$\begin{bmatrix} 1 & 0 & 1 \\ 4 & 4 & 3 \\ -4 & -3 & -3 \end{bmatrix} \begin{bmatrix} -3 & -3 & -4 \\ 0 & 1 & 1 \\ 4 & 3 & 4 \end{bmatrix} \mathbf{x} = \mathbf{x} = \begin{bmatrix} 1 & 0 & 1 \\ 4 & 4 & 3 \\ -4 & -3 & -3 \end{bmatrix} \begin{bmatrix} 21 \\ 21 \\ 16 \end{bmatrix}$$

$$= \begin{bmatrix} 37 \\ 216 \\ -195 \end{bmatrix}.$$

So, we decode $\begin{bmatrix} 24 & 17 & 6 \end{bmatrix}^T$ by computing

$$\begin{bmatrix} 1 & 0 & 1 \\ 4 & 4 & 3 \\ -4 & -3 & -3 \end{bmatrix} \begin{bmatrix} 24 \\ 17 \\ 6 \end{bmatrix} = \begin{bmatrix} 30 \\ 182 \\ -165 \end{bmatrix},$$

which becomes $\begin{bmatrix} 4 & 0 & 17 \end{bmatrix}^T$ modulo 26.

Not every matrix has an inverse. For this method to work, your matrix must have an inverse. Else, you can encode but can't decode. So, now you can pass notes and even encode them with linear algebra. Keep in mind, however, that even fancier methods are used to encode credit card numbers on the Internet as even this method can be broken quickly enough to be insecure for Internet transactions.

So, now we have a method of solving $A\mathbf{x} = \mathbf{b}$. We can use elementary row operations to compute such a solution and know how to apply this to encrypting data. When we solve a linear system like $A\mathbf{x} = \mathbf{b}$, we've computed an exact solution to the linear equations. In the next chapter, we see what can be done when there isn't a solution. In such cases, we'll find a $\mathbf{x}$ that, at least in a mathematical sense, comes close to approximating a solution.

SUPERCOMPUTING AND LINEAR SYSTEMS

The U. S. national laboratories house many of the world's fastest high performance computers. One of the fastest computers in the world (in fact the fastest in June 2010) is Jaguar (pictured above) housed at Oak Ridge National Laboratory that can perform 2.3 quadrillion floating point operations per second. Such computers are commonly used to solve large linear systems, sometimes containing billions of variables, which necessitates the use of large-scale parallel computing. These massive linear systems often result from differential equations that describe complex physical phenomena. *Image courtesy of Oak Ridge National Laboratory, U.S. Dept. of Energy*

7
Math to the Max

Sometimes, life is not linear. In this chapter, we'll model problems as linear processes, which will lead to linear systems such as $A\mathbf{x} = \mathbf{b}$. Since our data won't be exactly linear, a solution won't always exist to such a system. When a solution doesn't exist, there does not exist a vector $\mathbf{x}$ such that $A\mathbf{x}$ equals $\mathbf{b}$. In other words, if we could find such an $\mathbf{x}$ then $A\mathbf{x} - \mathbf{b} = \mathbf{0}$, which is the zero vector. Since we cannot achieve the zero vector by choosing $\mathbf{x}$ appropriately, our task will be to minimize the length of the vector $A\mathbf{x} - \mathbf{b}$. We apply this technique first in a quest to approximate when the fastest 100 meter race might be run, and later we look at the rankings for the best colleges in the United States.

7.1 Dash of Math

In 2012, Usain Bolt electrified the Olympic track and field stadium in London as he won a second consecutive gold medal in the 100 meter dash. This was the fastest time to date ever in the Olympics. No previous medalist could have beat him.

There are 28 gold medal times for the men's 100 m race in the Olympic Games between 1896 and 2012 with the times listed in Table 7.1. The slowest time was Tom Burke's 12 second sprint to gold in 1896. Bolt was the fastest in 2012. Let's get a sense of all the times by graphing them as seen in Figure 7.1 (a). We see an overall trend of decreasing times. It's a trend, so it may not always be true. In 1968, Jim Hines won gold in 9.95, which was the first sub-10 second time. It wouldn't be until 1984 and Carl Lewis that another gold would be won with a race under 10 seconds.

There is another trend in the data. It can be approximated by a line. While the line won't pass through every point, it can get close. Moreover,

Table 7.1. *Olympic Gold Medal Winning Times in Seconds for the Men's 100 Meter Dash.*

Year	Time	Year	Time	Year	Time	Year	Time
1896	12	1924	10.6	1960	10.32	1988	9.92
1900	11	1928	10.8	1964	10.06	1992	9.96
1904	11	1932	10.3	1968	9.95	1996	9.84
1906	11.2	1936	10.3	1972	10.14	2000	9.87
1908	10.8	1948	10.3	1976	10.06	2004	9.85
1912	10.8	1952	10.79	1980	10.25	2008	9.69
1920	10.8	1956	10.62	1984	9.99	2012	9.63

if the trend in the data continues, we can make predictions at how fast we might be running in 2020 or 2040. First, let's see how to model this data with a line and then how to predict the future.

We want to find a line $y = ax + b$, where x is a year and y is the time run in that year. If the points all lie along a line, then we'd have $9.63 = m(2012) + b$, using Bolt's second gold medal time. Similarly, Tom Burke's performance leads to the equation $12 = m(1896) + b$. The two linear equations, for Bolt and Burke, correspond to the system of equations

$$\begin{bmatrix} 2012 & 1 \\ 1896 & 1 \end{bmatrix} \begin{bmatrix} m \\ b \end{bmatrix} = \begin{bmatrix} 9.63 \\ 12 \end{bmatrix}.$$

We can easily solve this system and find $m = -0.0204$ and $b = 50.7372$. So, using only two times, we find the equation $y = -0.0204x + 50.7372$. Let's check this on times we didn't use to find the equation. For example, consider Carl Lewis's 1984 time of 9.99. We check the line by letting $x = 1984$. So, $y = -0.0204(1984) + 50.7372 = 10.2636$. The time doesn't match but it

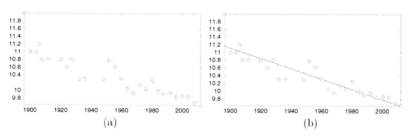

Figure 7.1. Olympic gold medal times in the men's 100 m dash (a) and a least squares fit to the data (b).

won't since (1984, 9.99), (2012, 9.63), and (1896, 12) will never lie on the same line. But, can we produce a better estimate with another line? Maybe we could try connecting another two points, but this takes considerable fiddling and isn't necessary. And, how would we know when we have the best line not just to estimate Carl Lewis's time but everyone's?

This issue leads us to the method of linear least squares. Let's add Carl Lewis's time to see, on a small system, how to find such a line that fits the data. We are now interested in the linear system

$$\begin{bmatrix} 2012 & 1 \\ 1896 & 1 \\ 1984 & 1 \end{bmatrix} \begin{bmatrix} m \\ b \end{bmatrix} = \begin{bmatrix} 9.63 \\ 12 \\ 9.99 \end{bmatrix},$$

which we will denote as $Mx = b$. This system doesn't have a solution. If it did, then we could find a line passing through all the points. So, instead, we want a line that is closest to all the points, possibly not passing through any.

Take a moment and consider the task. We have infinitely many choices for m and b. Let's pick two sets of values. Earlier, when we used only the times of Tom Burke and Usain Bolt, we found $m = -0.0204$ and $b = 50.7372$. These values correlate to a model in which the gold medal winning time decreases by 0.0204 of a second per year. The y-intercept of the line is how fast people, again under this model, would have run 100 meters in year 0. Is this model stating that the fastest person two thousand years ago would have run 100 meters in 50 seconds? In a sense, yes. Possibly a better way to look at it, though, is that the model doesn't accurately extend back that far. Keep in mind that our data covered just over one hundred years. Expecting it to accurately model phenomenon for two thousand years is asking too much.

Now, let's make another choice, and set $m = -0.021$ and $b = 51$. Remember, these values predict the values in our linear system, which are those gold medal winning times. To see what $m = -0.021$ and $b = 51$ predict for gold medal times, we compute

$$\begin{bmatrix} 2012 & 1 \\ 1896 & 1 \\ 1984 & 1 \end{bmatrix} \begin{bmatrix} -0.021 \\ 51 \end{bmatrix} = \begin{bmatrix} 8.7480 \\ 11.1840 \\ 9.3360 \end{bmatrix}.$$

The real winning times were

$$\begin{bmatrix} 9.63 \\ 12 \\ 9.99 \end{bmatrix}.$$

The error in the approximation is computed by

$$
\begin{bmatrix} 8.7480 \\ 11.1840 \\ 9.3360 \end{bmatrix} - \begin{bmatrix} 9.63 \\ 12 \\ 9.99 \end{bmatrix} = \begin{bmatrix} -0.8820 \\ -0.8160 \\ -0.6540 \end{bmatrix},
$$

which is called the *residual vector*.

Returning to our first choice of $m = -0.0204$ and $b = 50.7372$, the residual vector is

$$
\begin{bmatrix} 2012 & 1 \\ 1896 & 1 \\ 1984 & 1 \end{bmatrix} \begin{bmatrix} -0.0204 \\ 50.7372 \end{bmatrix} - \begin{bmatrix} 9.63 \\ 12 \\ 9.99 \end{bmatrix} = \begin{bmatrix} 9.6924 \\ 12.0588 \\ 10.2636 \end{bmatrix} - \begin{bmatrix} 9.63 \\ 12 \\ 9.99 \end{bmatrix}
$$

$$
= \begin{bmatrix} 0.0624 \\ 0.0588 \\ 0.2736 \end{bmatrix}.
$$

Between these two choices for m and b, which is better? To answer this, we need a mathematical sense of what is meant by "better." In this chapter, a smaller *least-squares error* implies a better approximation to the data. Least-squares error equals the sum of the squares of the terms of the residual vector.

For the residual vector

$$
\begin{bmatrix} -0.8820 \\ -0.8160 \\ -0.6540 \end{bmatrix}
$$

the least squares error equals $(-0.8820)^2 + (-0.8160)^2 + (-0.6540)^2 = 1.8715$. For the residual vector

$$
\begin{bmatrix} 0.0624 \\ 0.0588 \\ 0.2736 \end{bmatrix}
$$

the least squares error equals $(0.0624)^2 + (0.0588)^2 + (0.2736)^2 = 0.0822$. Since our first choice for m and b produced a lower least-squares error, the resulting line, $y = -0.0204x + 50.7372$, is a better fit to the data.

This is where the process of minimization enters. Our question is: among all choices for m and b, which produces the smallest least-squares error? The least-squares error is the square of the length of the residual vector. So, our problem can also be viewed as choosing values of m and b that produce the shortest residual vector.

How do we pick among an infinite number of choices for m and b? Somewhat amazingly, we need only to multiply both sides of the equation by M^T, which simply means we want to solve $M^T M \mathbf{x} = M^T \mathbf{b}$. For our small system, this corresponds to solving

$$\begin{bmatrix} 2012 & 1896 & 1984 \\ 1 & 1 & 1 \end{bmatrix} \begin{bmatrix} 2012 & 1 \\ 1896 & 1 \\ 1984 & 1 \end{bmatrix} \begin{bmatrix} m \\ b \end{bmatrix} = \begin{bmatrix} 2012 & 1896 & 1984 \\ 1 & 1 & 1 \end{bmatrix} \begin{bmatrix} 9.63 \\ 12 \\ 9.99 \end{bmatrix},$$

which, after performing the matrix multiplications, becomes

$$\begin{bmatrix} 106998484 & 54726 \\ 54726 & 28 \end{bmatrix} \begin{bmatrix} m \\ b \end{bmatrix} = \begin{bmatrix} 567962.44 \\ 290.84 \end{bmatrix},$$

or $m = -0.0210$ and $b = 51.8033$.

Let's revisit what makes this the best line. At $x = 2012$, it estimates the time to be $y = 2012(-0.0210) + 51.8033 = 9.5513$. The vertical distance from the point $(2012, 9.63)$ to the line is the distance between $(2012, 9.5513)$ and $(2012, 9.63)$, which equals $9.63 - 9.5513 = 0.0787$. For the year 1896, the line estimates the time as 11.9873, so the error is $12 - 11.9873 = 0.0127$. Finally, for 1984, the line would return $y = 10.1393$, and $10.1393 - 9.99 = 0.1493$.

The sum of the squares of the distances is $(0.0787)^2 + (0.0127)^2 + (0.1493)^2 = 0.0286$. This line, produced using the method of least squares, chose the m and b such that this sum is as small as possible. In this sense, it is the best fit of the data. No other line will produce a smaller sum.

What if we use all the data? This produces the line $y = -0.0133x + 36.31$, which is graphed in Figure 7.1. Notice how close the line is to various points. For 2012, it predicts Usain Bolt's time at 9.599, which is only .003 seconds off his actual time!

Let's look at our least squares line more closely. Its slope is -0.0133. Remember, x denotes the year and y is the time. So, the slope predicts that every year, we expect the Olympic gold medal time to drop by just over a hundredth of a second.

At what point would someone reach the limit of speed? No one has ever run the race in 2 seconds. So, there is some limit between 2 and 9.63 seconds. The book, *The Perfection Point*, considers such a question. It is written by John Brenkus, the host of ESPN's Sports Science show. Brenkus analyzes four distinct phases of the race: reacting to the gun, getting out of the blocks, accelerating to top speed, and hanging on for dear life at the end.

Table 7.2. *Statistics Reported by Three (fictional) Colleges.*

College	Overall Score	Percent of Classes with Fewer Than 20 Students	Percent of Faculty That are Full-Time	Freshman Retention
Gaussian University	81	54	86	89
Escher College	79	41	92	86
Matrix Tech	84	63	81	94

His work predicts 8.99 seconds as the fastest 100 meter dash that can ever be run. When does our model predict this to happen? We simply need to solve $8.99 = -0.013263412x + 36.31048075$. (I'm keeping more decimal places for the slope and y-intercept than I did earlier.) Solving for x, I find that 2059.84 is the answer, predicting that we'll see the fastest possible 100 meter dash for the men in the Olympics in the 2060 games!

7.2 Linear Path to College

As we looked at Olympic data, we were able to graph the points in two dimensions. We saw that a line could approximate the data. Now, we turn to an example, adapted from [7], that contains data in a much higher dimension. We won't graph the data but will still use least squares to approximate the values. For this, we turn to the college rankings of *U.S. News & World Report*, which have been offered since 1983. The underlying algorithm is kept under wraps. But, much of the data and, of course, the final rankings are published. Can't we simply back-engineer the formula? Easy to say (or write) but not so easy to do, which is pretty clear since no one has (publicly, at least) done it.

To see some of the issues that make finding the algorithm formidable, let's consider rating three colleges Gaussian University, Escher College, and Matrix Tech. Suppose the schools reported data percent of classes with fewer than 20 students, percent of faculty that are full-time, and freshman retention, as contained in Table 7.2.

Could we hack this rating method? It depends, in part, on the underlying formula. Suppose Gaussian University's overall score of 81 resulted from computing $54A + 86B + 89C$, where A, B, and C are some weights. Then $41A + 92B + 86C$ produced the final score for Escher College and $63A + 81B + 94C$ the score for Matrix Tech. The system to find A, B and C is straightforward—a linear system,

$$\begin{bmatrix} 54 & 86 & 89 \\ 41 & 92 & 86 \\ 63 & 81 & 94 \end{bmatrix} \begin{bmatrix} A \\ B \\ C \end{bmatrix} = \begin{bmatrix} 81 \\ 79 \\ 84 \end{bmatrix}.$$

Solving, we find that

$$\begin{bmatrix} A \\ B \\ C \end{bmatrix} = \begin{bmatrix} 0.1165 \\ 0.2284 \\ 0.6187 \end{bmatrix}.$$

Is this the scaling used? In fact, no. The weights were

$$\begin{bmatrix} A \\ B \\ C \end{bmatrix} = \begin{bmatrix} 0.2 \\ 0.3 \\ 0.5 \end{bmatrix},$$

which were not recovered since the exact score was not reported. For instance, $0.2(54) + 0.3(86) + 0.5(89) = 81.1$, which was reported as 81. The rounding from 81.1 (computed value) to 81 (reported value) caused us to find different weights in our calculation. Still, this gave a good sense of the weights and especially their relative importance. For example, raising freshman retention by 1% will result in a higher overall score than raising the percent of classes with fewer than 20 students even by 2%.

Now, suppose another college, U2, reported data of 71, 79, and 89 for percent of classes with fewer than 20 students, percent of faculty that are full-time, and freshman retention, respectively. U2's score would be 82. Since the formula involves three unknowns A, B, and C, we need only the data for three schools, which we just used. But, we could use U2's data and overall score to check our method. Using our earlier computed weights, $0.1165(71) + 0.2284(79) + (0.6187)89 = 81.3813$, which affirms that our weights are close, although not exact.

Let's try this type of approach on the *U.S. News & World Report* data from 2013 for national liberal arts schools ranked in the top 25, found at the end of the chapter in Table 7.5. The data contains information on the college's overall score as measured on its academic reputation, selectivity rank, SAT (VM) 25th-75 percentile, percent of freshmen in top 10% of their high school class, acceptance rate, faculty resource rank, percent of classes with fewer than 20, percent of classes with greater than or equal to 50 students, student/faculty ratio, percent of faculty that are full-time, graduation retention rank, freshman retention, financial resources rank, and alumni giving rank. Excluding the columns for the score and name of the college, we will use the remaining columns in Table 7.5 to find a linear combination to create the score. The only column that is changed is that for SAT scores, which is reported as a range like 1310–1530 for Williams College. Since we will want one number per column, we will take the average of the range, which for Williams would be 1420.

Given there are fourteen columns of data, we need only the data from the top fourteen colleges that correspond to data from the rows for Williams College to the US Naval Academy. This creates the linear system

$$
\begin{bmatrix}
92 & 4 & 1420 & 91 & 17 & 3 & 71 & 4 & 7 & 93 & 1 & 97 & 6 & 58 \\
92 & 5 & 1425 & 84 & 13 & 7 & 70 & 2 & 9 & 94 & 1 & 98 & 10 & 57 \\
91 & 6 & 1440 & 84 & 15 & 7 & 74 & 2 & 8 & 93 & 4 & 97 & 9 & 46 \\
87 & 6 & 1385 & 86 & 18 & 17 & 68 & 1 & 9 & 94 & 11 & 96 & 3 & 55 \\
87 & 2 & 1460 & 90 & 14 & 20 & 70 & 1 & 8 & 94 & 1 & 98 & 6 & 43 \\
87 & 8 & 1410 & 83 & 16 & 14 & 68 & 1 & 10 & 93 & 6 & 96 & 14 & 50 \\
89 & 12 & 1390 & 78 & 31 & 12 & 69 & 1 & 8 & 93 & 14 & 95 & 10 & 46 \\
88 & 12 & 1415 & 78 & 31 & 16 & 65 & 1 & 9 & 97 & 4 & 97 & 27 & 58 \\
83 & 2 & 1400 & 94 & 25 & 5 & 79 & 1 & 8 & 94 & 6 & 96 & 15 & 44 \\
85 & 14 & 1390 & 71 & 14 & 4 & 86 & 2 & 9 & 94 & 11 & 96 & 21 & 43 \\
88 & 14 & 1395 & 74 & 23 & 20 & 68 & 0.3 & 8 & 95 & 6 & 97 & 13 & 33 \\
83 & 10 & 1360 & 82 & 28 & 15 & 69 & 0 & 11 & 99 & 6 & 96 & 37 & 53 \\
89 & 1 & 1500 & 95 & 22 & 18 & 67 & 2 & 8 & 97 & 21 & 98 & 18 & 33 \\
88 & 46 & 1270 & 53 & 7 & 24 & 61 & 0 & 9 & 94 & 25 & 97 & 1 & 21
\end{bmatrix}
\mathbf{w} =
\begin{bmatrix}
100 \\ 98 \\ 96 \\ 94 \\ 94 \\ 93 \\ 93 \\ 92 \\ 91 \\ 90 \\ 90 \\ 89 \\ 89 \\ 88
\end{bmatrix} .
$$

Easy enough to solve. Keep in mind, solving the system produces weights that compute the exact overall score for the fourteen schools. So, we now apply them to the remaining schools and see the predicted scores. Here, we see how poorly we've done. For example, Washington and Lee's overall score was 88; our method predicts 73.7. Wesleyan scored 86 although we predict 118. Also, Colby and Colgate both scored 84 and our method gives them scores of 88 and 96. So, we may seem at a loss. But, as is often the case in mathematics, meeting a dead end may mean that only a slight change in perspective will lead to a fruitful new direction.

Like the example with Olympic data, we weren't using all the data. Using least squares enabled us to include more gold medal times and can let us include more schools here. We'll only use the first 20 schools so we have a few left against which to test our method. This would produce the system

$$
D_{20 \times 14} \mathbf{w}_{14 \times 1} = \mathbf{s}_{20 \times 1},
$$

Table 7.3. *Predicted and Actual Scores for
Schools Using Linear Algebra and Least Squares.*

School	Predicted Score	Actual Score
Bates	83.5	83
Macalester	81.6	82
Bryn Mawr	83.6	81
Oberlin	79.2	81

where D is the data for the schools, $\mathbf{w}$ is the computed weights, and $\mathbf{s}$ is the final overall score. Trying to solve this system, we quickly see that we need least squares, as in the previous section.

Again, we'll omit some data from our linear system in order to see the accuracy of the predicted scores. When we have computed $\mathbf{w}$ for the linear system, we find the predicted scores for the schools omitted from the linear system; this produces Table 7.3. Grinnell was omitted since no data was submitted for the SAT. Now, look at the predicted scores and how closely they come to approximating the overall scores computed by *U.S. News & World Report*. Keep in mind, though, that we solved the system using least squares. So, now, our predicted scores even for schools in the original system are no longer exact. For instance, the predicted scores for Williams and Amherst were 99.5 and 98.2, whereas the actual overall scores were 100 and 98.

Let's reflect on our method. First, the least squares method for solving a linear system does a reasonable job at predicting the actual scores of colleges. Keep in mind that it is unlikely these are the actual weights since (1) the reported scores were probably rounded and (2) the scores were probably scaled to give that year's top college, in this case Williams, a score of 100.

Insights are possible, even with our inexact weights. Interestingly, academic reputation has the highest weight. It's over double the magnitude of the next largest weight.

This is interesting in light of a controversy that occurred in 2007 when the Presidents Letter was signed by twelve college and university presidents. It called for colleges and universities to no longer use the rankings as an indication of their quality and to refuse to fill out the reputational survey. Since the original letter was published, more than 50 other college and university presidents joined in signing the letter. The letter can be found on the website for The Education Conservancy.

What happens if we remove the highest weight and keep all the other weights the same? How does this affect rank? Are Williams and Amherst weakened dramatically if they lose weight and drop their reputation score? It is interesting that in 2013, the top five schools—Williams, Amherst, Swarthmore, Middlebury, and Pomona—retain the same ranking. But Haverford jumps from ninth to sixth. Keep in mind many more schools appear in the list. How would removing the reputation score impact your favorite college's rankings? Look online, find the data and begin asking your questions!

U.S. News & World Report's keeps its ranking formula a secret, yet the method of least squares enables us to use available data to approximate the underlying algorithm.

7.3 Going Cocoa for Math

We've been discussing the rankings of colleges. Now let's see how eating chocolate might help a math major in college, or at least how we can use least squares to make such an inference.

Claiming that eating chocolate helps with math isn't new. Researchers in the performance and nutrition research centre at Northumbria University found that mental arithmetic became easier after volunteers had been given large amounts of compounds found in chocolate, called flavanols, in a hot cocoa drink. Their findings suggest that binging on chocolate when preparing for an exam.

A paper in the *New England Journal of Medicine* in 2012 found a correlation between chocolate consumption and the number of Nobel Laureates produced. Countries with higher chocolate consumption were found to produce more Nobel Laureates. The article estimated that a rise of consumption of 0.4 kg per person per year might correlate with an increase of one Nobel Prize winner per year.

What connection might we make? Not surprisingly, we'll look to least squares as our tool of analysis. In particular, we'll look at the number of doctorates in engineering and science awarded in the United States and chocolate consumption (measured in trillions of dollars) in the United States as found in Table 7.4.

First, we plot the data. We'll take the x-axis to be dollars spent on chocolate. The y-axis will be the number doctorates awarded in science and engineering in the United States. As seen in the graph in Figure 7.2, the data is close to linear, with the least squares line being $y = 2560.6x + 880.8$.

Table 7.4. *The Number of Doctorates in Engineering and Science Awarded in the United States and Chocolate Consumption (measured in trillions of dollars) in the United States.*

Year	No. Doctorates	Chocolate Consumption
2002	24,608	9.037
2003	25,282	9.52
2004	26,274	10.13
2005	27,986	10.77
2006	29,863	11.39
2007	31,800	11.95
2008	32,827	12.37

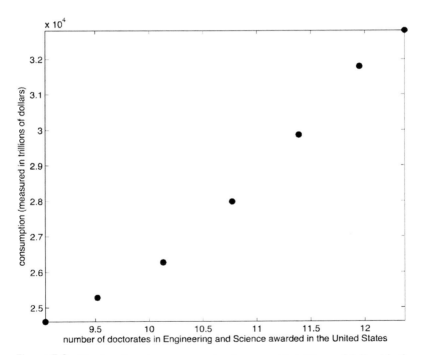

Figure 7.2. Plotting chocolate consumption (measured in trillions of dollars) in the United States versus the number of doctorates in engineering and science awarded in the United States.

This little equation contains a lot of information. It states that according to this data, under this linear model, for each trillion of dollars spent on chocolate, an additional 2,560 doctorates will be awarded in these fields! And, if we spend nothing on chocolate, then there will be 880 doctorates awarded!

To further this argument, let's see how correlated this data is. To compute correlation, we first subtract the mean from each column of data in Table 7.4. If we let **c** and **d** equal the mean-subtracted vectors of data, then the correlation equals

$$\text{corr}(\mathbf{c}, \mathbf{d}) = \frac{\mathbf{c} \cdot \mathbf{d}}{\|\mathbf{c}\| \|\mathbf{d}\|},$$

which we saw earlier with the dot product. Indeed, we are again measuring an angle, which is helping us find correlation. If $\text{corr}(\mathbf{c}, \mathbf{d}) = 1$, then the vectors **c** and **d** point in the same direction and the data is perfectly correlated— knowing the value of one quantity enables you to exactly compute the value of the other. For this data, the correlation coefficient equals 0.99. We have found strong correlation between doctorates in science and engineering and chocolate consumption.

Ready to ask the government to invest in chocolate so we see more doctorates in science and engineering? Before calling your legislator, keep in mind an important saying in statistics,

CORRELATION DOESN'T MEAN CAUSATION.

For example, it's been shown that shark attacks are correlated with ice cream consumption. Consider a few more correlations found by Harvard law student Tyler Vigen on his website Spurious Correlations [16]. Sour cream consumption correlates with motorcycle accidents. He also found that the number of people who drown after falling out of a fishing boat correlates with the marriage rate.

Did you find yourself trying to think why these events might be connected? That's exactly because we tend to think about correlation being linked with causation, which can give us keen insight but also misleading intuition.

Correlations can be real. Causation can lead to correlation. But that doesn't mean that correlation means causation. Two factors may be correlated due to their association with another factor. We may simply be seeing a random connection in data. Even when we see a strong and repeatable

Table 7.5. *Data Published for U.S. News & World Report 2013 Ranking. The Columns Correspond to the Following Categories, Respectively: College, Overall Score, Academic Reputation, Selectivity Rank, SAT (VM) 25th–75th Percentile, Percent Freshmen in Top 10% of HS Class, Acceptance Rate, Faculty Resource Rank, Percent of Classes with Fewer Than 20, Percent of Classes with More Than 50 Students, Student/Faculty Ratio, % Faculty Who are Full-Time, Graduation Retention Rank, Freshman Retention, Financial Resources Rank, Alumni Giving Rank.*

College	Score	Aca.	Sel	SAT	Top 10	Acc.	Fa. R.	< 20	≥ 50	Rat.	% FT	Gr. R.	Fr. R.	Fin.	Alum.
Williams	100	92	4	1310–1530	91	17	3	71	4	7	93	1	97	6	58
Amherst	98	92	5	1320–1530	84	13	7	70	2	9	94	1	98	10	57
Swarthmore	96	91	6	1350–1530	84	15	7	74	2	8	93	4	97	9	46
Middlebury	94	87	6	1290–1480	86	18	17	68	1	9	94	11	96	3	55
Pomona	94	87	2	1370–1550	90	14	20	70	1	8	94	1	98	6	43
Bowdoin	93	87	8	1330–1490	83	16	14	68	1	10	93	6	96	14	50
Wellesley	93	89	12	1290–1490	78	31	12	69	1	8	93	14	95	10	46
Carleton	92	88	12	1320–1510	78	31	16	65	1	9	97	4	97	27	58
Haverford	91	83	2	1300–1500	94	25	5	79	1	8	94	6	96	15	44
Claremont McKenna	90	85	14	1300–1480	71	14	4	86	2	9	94	11	96	21	43
Vassar	90	88	14	1320–1470	74	23	20	68	0.3	8	95	6	97	13	33
Davidson	89	83	10	1270–1450	82	28	15	69	0	11	99	6	96	37	53
Harvey Mudd	89	89	1	1430–1570	95	22	18	67	2	8	97	21	98	18	33
US Naval Academy	88	88	46	1160–1380	53	7	24	61	0	9	94	25	97	1	21
Washington and Lee	88	78	9	1310–1480	81	18	2	74	0.2	9	91	14	94	25	46
Hamilton	87	81	17	1310–1470	74	27	6	74	1	9	94	21	95	23	47
Wesleyan	86	85	19	1300–1480	66	24	48	68	5	9	97	6	96	29	49
Colby	84	81	25	1250–1420	61	29	20	69	2	10	93	14	95	29	41
Colgate	84	83	19	1260–1440	67	29	29	64	2	9	95	14	94	32	40
Smith	84	85	35	1200–1440	61	45	20	66	5	9	97	35	92	21	36
US Military Academy	83	83	28	1150–1370	58	27	43	67	3	10	95	14	94	35	46
Bates	83	83	28	1260–1420	58	27	43	67	3	10	95	14	94	35	46
Grinnell	83	86	32	NA	62	51	29	62	0.3	9	91	28	94	27	40
Macalester	82	83	19	1240–1440	70	35	35	70	1	10	89	28	94	41	39
Bryn Mawr	81	83	39	1200–1430	60	46	27	74	3	8	90	47	92	23	40
Oberlin	81	82	19	1280–1460	68	30	43	70	3	9	96	32	94	37	38

correlation, we may not have found that change in one variable causes a change in another.

As is often the case, math can be a powerful tool. Yet, one must wield it carefully and not overestimate its abilities. If we see that two phenomenon are correlated, we do not know why or that one causes the other. In a similar way, finding a least squares fit to a set of data may approximate the data but it does not make a definitive claim about the future. It may but dynamics can change. For instance, will Olympic runners continue to decrease their times at the rate we've seen? Possibly. But, we may also find that the rates plateau and possibly continue to decrease, but at a much slower rate. Time will tell. Either way, the mathematics of this section aided us in understanding data and gaining insights that sometimes foster more questions.

Between the last and current chapters, we've looked at solving $A\mathbf{x} = \mathbf{b}$. As we proceed through the remaining chapters, we'll continue to explore applications of algebra with matrices. We'll also see that linear algebra involves much more than solving a linear system. For example, in the next chapter, we discuss a special type of vector and later, when we discuss data analytics in more depth, we'll see how computing this type of vector in the context of modeling the World Wide Web helped Google become a billion dollar business.

8
Stretch and Shrink

Multiplying a vector $\mathbf{x}$ by a matrix A results in another vector, possibly in a different dimension. Take the 3D vector $\begin{bmatrix} 1 & 3 & -2 \end{bmatrix}^T$ and use it in the product

$$\begin{bmatrix} 1 & 2 & -1 \\ 3 & 0 & 1 \end{bmatrix} \begin{bmatrix} 1 \\ 3 \\ -2 \end{bmatrix}.$$

This results in the vector

$$\begin{bmatrix} 9 \\ 1 \end{bmatrix},$$

which is in 2D. Some matrices, when multiplied by a vector, will stretch or shrink the vector. For example,

$$\begin{bmatrix} 2 & -4 \\ -1 & -1 \end{bmatrix} \begin{bmatrix} -4 \\ 1 \end{bmatrix} = \begin{bmatrix} -12 \\ 3 \end{bmatrix} = 3 \begin{bmatrix} -4 \\ 1 \end{bmatrix}.$$

These types of vectors result in an equation of the form $A\mathbf{v} = \lambda\mathbf{v}$. Said in words, this type of vector $\mathbf{v}$ has the property that the matrix A multiplied by $\mathbf{v}$ results in a scalar multiple of $\mathbf{v}$. In such cases, $\mathbf{v}$ is called an *eigenvector* of A and λ its associated *eigenvalue*. The terms eigenvalue and eigenvector are derived from the German word "Eigenwert" which means "proper value." The word eigen is pronounced "eye-gun."

8.1 Getting Some Definition

Let's take a closer look at an eigenvector and eigenvalue. We have $A\mathbf{v} = \lambda\mathbf{v}$, which indicates that λ is an eigenvalue associated with its corresponding eigenvector **v**. To see what's involved, let's look at 2×2 matrices. The determinant of a 2×2 matrix

$$\det(A) = \begin{bmatrix} a_{11} & a_{12} \\ a_{21} & a_{22} \end{bmatrix}$$

is

$$\det(A) = \begin{vmatrix} a_{11} & a_{12} \\ a_{21} & a_{22} \end{vmatrix} = a_{11}a_{22} - a_{21}a_{12}.$$

So,

$$\begin{vmatrix} 1 & 2 \\ 4 & 3 \end{vmatrix} = (1)(3) - (4)(2) = 3 - 8 = -5.$$

The equation $A\mathbf{v} = \lambda\mathbf{v}$ can be rewritten as $A\mathbf{v} - \lambda\mathbf{v} = \mathbf{0}$ or $(A - \lambda I)\mathbf{v} = \mathbf{0}$. Such a nonzero vector **v** will exist if $\det(A - \lambda I) = 0$. This tells us how to find the eigenvalue. If

$$A = \begin{bmatrix} 1 & 2 \\ 4 & 3 \end{bmatrix}$$

then we want to find values of λ such that

$$\det(A - \lambda I) = \begin{vmatrix} 1 - \lambda & 2 \\ 4 & 3 - \lambda \end{vmatrix} = 0.$$

This happens when $(1 - \lambda)(3 - \lambda) - 8 = \lambda^2 - 4\lambda - 5 = 0$. Solving the quadratic equation leads to $\lambda = -1$ or $\lambda = 5$.

We just found the roots of the characteristic polynomial. For a 2×2 matrix, we find the roots of a quadratic to find the eigenvalues. Finding the eigenvalues of an $n \times n$ matrix involves computing roots of a polynomial of degree n. Once you find an eigenvalue, you substitute it into $(A - \lambda I)\mathbf{v} = \mathbf{0}$ and use Gaussian elimination to find **v**. If you have a matrix of degree 10, 20, or 100, finding the roots of the characteristic polynomials can be quite difficult. So, computers often estimate the eigenvalues and eigenvectors of a matrix. For the remainder of the chapter, we'll assume we have a matrix calculator that has eigenvector and eigenvalue buttons. Many scientific calculators have the ability to make such calculations.

(a) (b) (c)

Figure 8.1. Shearing the Mona Lisa.

8.2 Getting Graphic

What's happening visually with an eigenvector? One application is a *shear mapping* of an image, which leaves points along one axis unchanged while shifting other points in a direction parallel to that axis by an amount that is proportional to their perpendicular distance from the axis. For example, a horizontal shear in the plane is performed by multiplying vectors by the matrix

$$S = \begin{bmatrix} 1 & k \\ 0 & 1 \end{bmatrix}.$$

Using this matrix, the vector $[x\ y]^T$ is mapped to $[x + ky\ y]^T$, where $k = \cot\phi$ and ϕ is the angle of a sheared square to the x-axis. For example, a shear mapping of the image of the Mona Lisa in Figure 8.1 (a) produces the image in Figure 8.1 (b).

The vector $[1\ 0]^T$ is an eigenvector of the matrix S. The mapping leaves points along the x-axis unchanged but shifts all others. Repeatedly applying a shear transformation changes the direction of all vectors in the plane closer to the direction of the eigenvector as can be seen in Figure 8.1 as (b) is produced by a shear mapping of (a) and the same shear mapping is applied to the image in (b) to produce (c).

As another example, consider a rubber sheet that is stretched equally in all directions. This means that all vectors $[x\ y]^T$ in the plane are multiplied by the same scalar λ; that is, vectors are multiplied by the *diagonal* matrix

$$\begin{bmatrix} \lambda & 0 \\ 0 & \lambda \end{bmatrix}.$$

(a)   (b)

Figure 8.2. An original image is repeatedly scaled via a linear transformation.

This matrix's only eigenvalue is λ, and any 2D column matrix is an eigenvector. Figure 8.2 shows the effect of repeatedly applying this transformation with $\lambda = 1/3$ in (a) and with $\lambda = -1/3$ in (b). In both figures, we have placed the original image and the images produced by three applications of this transformation on top of each other. Knowing the values of λ, can you determine which of the four images in (a) and in (b) is the starting image?

8.3 Finding Groupies

Clustering is a powerful tool in data mining that sorts data into groups called *clusters*. Which athletes are similar, which customers might buy similar products, or what movies are similar are all possible topics of clustering. There are many ways to cluster data and the type of question you ask and data you analyze often influences which clustering method you will use. In this section, we'll cluster undirected graphs using eigenvectors.

We'll see how eigenvectors play a role in clustering by partitioning the undirected graph in Figure 8.3. A *graph* is a collection of vertices and edges. The vertices in this graph are the circles labeled 1 to 7. Edges exist between pairs of vertices. This graph is undirected since the edges have no

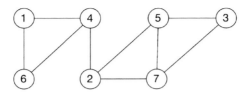

Figure 8.3. An undirected graph to cluster.

orientation. Think of the edges as connections. We see that 1 is connected to 4 and since the graph is undirected that means 4 is also connected to 1. In a directed graph, there is orientation. So, 1 may be connected to 4 but 4 may not be connected to 1. In such graphs, this is denoted by an arrow.

To cluster this graph, we'll use the eigenvector of a matrix. First, we need to form the matrix, which begins by forming the adjacency matrix A of the graph where $a_{ij} = 1$ if there is an edge between nodes i and j and is 0 otherwise. So, for Figure 8.3,

$$A = \begin{bmatrix} 0 & 0 & 0 & 1 & 0 & 1 & 0 \\ 0 & 0 & 0 & 1 & 1 & 0 & 1 \\ 0 & 0 & 0 & 0 & 1 & 0 & 1 \\ 1 & 1 & 0 & 0 & 0 & 1 & 0 \\ 0 & 1 & 1 & 0 & 0 & 0 & 1 \\ 1 & 0 & 0 & 1 & 0 & 0 & 0 \\ 0 & 1 & 1 & 0 & 1 & 0 & 0 \end{bmatrix}.$$

The next step is forming the matrix D which is a diagonal matrix where d_{ii} equals the row sum of the ith row of the matrix A. Since the first row of A sums to 2, the diagonal element of D in the first row is set to 2. Similarly, since the second row of A sums to 3, the diagonal element in that row of D equals 3. Continuing in this way, we find

$$D = \begin{bmatrix} 2 & 0 & 0 & 0 & 0 & 0 & 0 \\ 0 & 3 & 0 & 0 & 0 & 0 & 0 \\ 0 & 0 & 2 & 0 & 0 & 0 & 0 \\ 0 & 0 & 0 & 3 & 0 & 0 & 0 \\ 0 & 0 & 0 & 0 & 3 & 0 & 0 \\ 0 & 0 & 0 & 0 & 0 & 2 & 0 \\ 0 & 0 & 0 & 0 & 0 & 0 & 3 \end{bmatrix}.$$

We now can form the matrix we need

$$L = D - A = \begin{bmatrix} 2 & 0 & 0 & 0 & 0 & 0 & 0 \\ 0 & 3 & 0 & 0 & 0 & 0 & 0 \\ 0 & 0 & 2 & 0 & 0 & 0 & 0 \\ 0 & 0 & 0 & 3 & 0 & 0 & 0 \\ 0 & 0 & 0 & 0 & 3 & 0 & 0 \\ 0 & 0 & 0 & 0 & 0 & 2 & 0 \\ 0 & 0 & 0 & 0 & 0 & 0 & 3 \end{bmatrix} - \begin{bmatrix} 0 & 0 & 0 & 1 & 0 & 1 & 0 \\ 0 & 0 & 0 & 1 & 1 & 0 & 1 \\ 0 & 0 & 0 & 0 & 1 & 0 & 1 \\ 1 & 1 & 0 & 0 & 0 & 1 & 0 \\ 0 & 1 & 1 & 0 & 0 & 0 & 1 \\ 1 & 0 & 0 & 1 & 0 & 0 & 0 \\ 0 & 1 & 1 & 0 & 1 & 0 & 0 \end{bmatrix}$$

$$= \begin{bmatrix} 2 & 0 & 0 & -1 & 0 & -1 & 0 \\ 0 & 3 & 0 & -1 & -1 & 0 & -1 \\ 0 & 0 & 2 & 0 & -1 & 0 & -1 \\ -1 & -1 & 0 & 3 & 0 & -1 & 0 \\ 0 & -1 & -1 & 0 & 3 & 0 & -1 \\ -1 & 0 & 0 & -1 & 0 & 2 & 0 \\ 0 & -1 & -1 & 0 & -1 & 0 & 3 \end{bmatrix}.$$

This matrix is called the *Laplacian*.

Identifying this matrix's role in clustering is due to the work of Miroslav Fiedler. He uncovered the importance of the eigenvector corresponding to the second smallest eigenvalue of the Laplacian matrix. In particular, he proved that this vector will partition a graph into maximally intraconnected components and minimally interconnected components. This fits what we want with clusters. Think of a close group of friends in a class. They have strong friendships in their clique but fewer connections outside the group. The importance of the vector Fiedler identified led to it being referred to as the *Fiedler vector*.

What clusters does the Fiedler vector identify for Figure 8.3? Given how small the graph is, you may want to make a guess before letting the mathematics inform you. Computing the eigenvector associated with the second smallest eigenvalue, we find the Fiedler vector is

$$\begin{bmatrix} 0.4801 \\ -0.1471 \\ -0.4244 \\ 0.3078 \\ -0.3482 \\ 0.4801 \\ -0.3482 \end{bmatrix}.$$

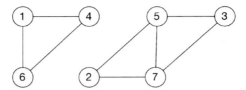

Figure 8.4. A clustering of the undirected graph in Figure 8.3.

Clustering is now quite simple. The Fiedler method clusters the graph using the signs of the eigenvector. The rows with the same sign are placed in the same cluster. So our previous vector, looking only at signs is

$$\begin{bmatrix} + \\ - \\ - \\ + \\ - \\ + \\ - \end{bmatrix}.$$

So, we place nodes 1, 4, and 6 into one cluster and nodes 2, 3, 5, and 7 into the other. Removing the edges (which is only the edge between nodes 2 and 4) between clusters, we get the clusters seen in Figure 8.4.

Let's try this on a larger network. We'll analyze some Facebook friends. We'll use the social network of a friend who joined the social network recently. In the undirected graph, the nodes will be the Facebook friends. There is an edge between two people (nodes) if they are friends on Facebook. We see the graph of the close to 50 people in Figure 8.5 (a). Again, the goal of clustering is to form maximally connected groups while the connections between different groups are minimally connected. How to do this is not immediately obvious in Figure 8.5 (a). Overall, we see little organization or a pattern of connectivity within the group. Let's partition the group into two clusters using the Fiedler method. If we color the nodes and place them close to each other, we see the graph in Figure 8.5 (b).

There are still connections between clusters. This is due to friendships that cross the two groups. For each cluster, can we identify attributes that describe its members? Such insight isn't outputted by the Fiedler method. I asked my friend to look at the clusters in Figure 8.5 (b). The connected graph with black nodes (to the right) is my friend's extended family, which, in part, explains the large number of connections within the network. The other

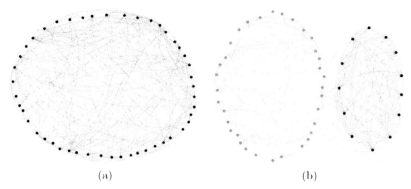

(a)   (b)

Figure 8.5. Clustering approximately 50 people by their friendships on Facebook. The network (a) and when grouped by the clusters created using one eigenvector (b).

group is essentially non-family members although a few family members make their way into this cluster.

Plotting a matrix is another useful way to view the effect of clustering, especially if the network contains hundreds or thousands of nodes. The plot of the adjacency matrix of the graph in Figure 8.5 (a) appears in Figure 8.6 (a). After clustering, we reorder the matrix so rows in a cluster appear together in the matrix. Having done this, we see the matrix in Figure 8.6 (b).

The goal of clustering is to form groups with maximally connected components within a group and minimally connected components between different groups. We see this with the smaller, but dense, submatrix in the upper left and the larger submatrix in the lower right. The darker square regions in the plot of the matrix correspond to the clusters in Figure 8.5 (b). What if we tried more clusters? Can we? We will learn to break the matrix into more groups in Chapter 11.

8.4 Seeing the Principal

In this section, we see how eigenvectors can aid in data reduction and unveil underlying structure in information. As an example adapted from [11], let's look at data sampled from an ellipse centered at the origin as seen in Figure 8.7. We are looking for what are called the *principal components*. We'll use the method called *Principal Component Analysis* or PCA.

For this example, we'll find one principal component. It is the direction where there is the most variance. First, let's find the variance in the horizontal direction. We compute the square of the distance from every point to

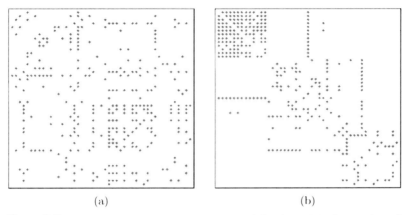

(a)   (b)

Figure 8.6. Looking at the Facebook social network in Figure 8.5 of approximately 50 friends via the adjacency matrix.

the x-axis, illustrated by the dotted lines in Figure 8.8 (a). This equals 600, giving the variance of the data over the y-coordinates. That is, the measure of how spread out is the data in the y-direction. Compare this to the variance in the vertical direction. The square of the distance from every point to the y-axis, illustrated by the dotted lines in Figure 8.8 (b), is 2400. From these two numbers, we find the data to be more spread out over the x-coordinates. You can see this visually since the data points are farther away from each other horizontally than vertically.

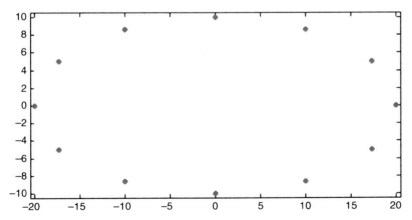

Figure 8.7. Points sampled from an ellipse centered at the origin.

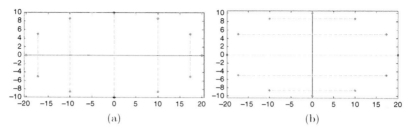

Figure 8.8. Finding the variance of the points in Figure 8.7 from the y-axis (a) and x-axis (b).

The most important point is that no other straight line has a larger variance than the vertical one. So, the largest variance of the data occurs over the x-coordinates. Therefore, the x-axis is the principal component for this data. How do I know no better line exists? An eigenvector told me. Let's see how.

For eigenvectors to help us, we first construct a matrix whose columns contain the coordinates of the points in our data. For our example, our matrix of points in Figure 8.7, rounded to the first decimal place, is

$$P = \begin{bmatrix} 20 & 17.3 & 10.0 & 0 & -10.0 & -17.3 & -20 & -17.3 & -10.0 & 0 & 10.0 & 17.3 \\ 0 & 5.0 & 8.7 & 10 & 8.7 & 5.0 & 0 & -5.0 & -8.7 & -10 & -8.7 & -5.0 \end{bmatrix}.$$

The matrix that we analyze is $C = PP^T$, which, for this example, is

$$\begin{bmatrix} 2400 & 0 \\ 0 & 600 \end{bmatrix}.$$

For a diagonal matrix the eigenvalues are the diagonal elements. The eigenvalues indicate the variance in the direction of the corresponding eigenvector. So, the largest eigenvalue is 2400 and its associated eigenvector is:

$$\begin{bmatrix} 1 \\ 0 \end{bmatrix},$$

which matches what we found earlier.

Now let's rotate the data by 50 degrees as seen in Figure 8.9 (a). If we again form the matrix $C = PP^T$, where the points are now the rotated points seen in Figure 8.9 (a) then the eigenvalues are again 2400 and 600 but now the eigenvector associated with the largest eigenvalue is

$$\begin{bmatrix} -0.6428 \\ 0.7660 \end{bmatrix}.$$

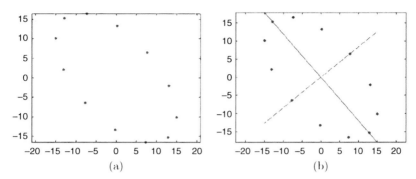

Figure 8.9. Points in Figure 8.7 rotated (a). The first and second principal components (b).

This vector, indeed, passes through the data at the same spot as the x-axis in the data in Figure 8.7. Our eigenvectors indicate the amount of variance of the data in that direction. So, the eigenvector with the highest eigenvalue is the principal component. The second eigenvector is perpendicular to the first and reframes the data in a new coordinate system as seen in Figure 8.9 (b).

One of the biggest advantages of this method is in dimension reduction. Suppose we rotate the ellipse into 3D as seen in Figure 8.10. The data is still two dimensional in nature. It has length and width but no height which is why it all lies in a plane, which is drawn with a dotted line in Figure 8.10. For this type of data, we will find three eigenvectors but one of them will have an eigenvalue of zero indicating the data actually is two dimensional.

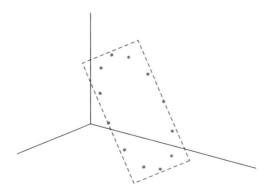

Figure 8.10. A 2D ellipse rotated in 3D.

So, we can again reframe the data but in two dimensions. This could be extended to 4, 5, or 10 dimensions with data living in a 2D plane in that higher dimension. As such, we can take data in 10 dimensions or even 100 and reduce it down to two dimensions.

We can reduce dimensions even if there isn't a zero eigenvalue. Imagine a slight variation to the example: instead of the oval being on a 2D plane, there is a perturbation of the coordinates above and below the plane. For this type of problem, there would still be three eigenvectors, but none would equal zero. The values might be something like 10, 8, and 0.1. The eigenvalue of 0.1 indicates that there is very little variation in the data in the direction of the corresponding eigenvector. Earlier, we saw that all the information existed in two dimensions which resulted in an eigenvector with an eigenvalue of 0. We can discard this dimension and project the data into two dimensions, making the dataset simpler. The data could have existed in five or 50 dimensions and, if a similar dynamic exists, could have been brought down to two, simplifying the problem and reducing the size of the data.

Our 2D ellipse was centered at $(0, 0)$. If it were centered somewhere else, like $(5, 4)$, then we'd take all the x-coordinates and subtract their mean. Similarly, we'd subtract the mean for the y-coordinates. This centers the ellipse at the origin. For PCA, the data is always preprocessed to subtract the mean in such a way. In Chapter 11, we'll see how to apply this idea to facial recognition.

As we saw, PCA can reduce the dimensions of a problem. In doing so, we approximate data, sometimes reducing the noise in measurement or capturing the nature of the data. In the next chapter, we approximate data by assuming it is linear.

9

Zombie Math—Decomposing

Any positive number can be factored into its prime factors. For instance, $315 = 3^2(5)(7)$. We've *decomposed* the number 315 into its prime factors.

A matrix can also be decomposed. Given matrices A, M, and N, if

$$A = MN$$

then we say that the matrices M, and N form a *matrix decomposition* of A since their product equals A. For example, if

$$A = \begin{bmatrix} 1 & 2 \\ 3 & 4 \end{bmatrix}, \quad L = \begin{bmatrix} \frac{1}{3} & 1 \\ 1 & 0 \end{bmatrix} 0, \quad \text{and} \quad U = \begin{bmatrix} 3 & 4 \\ 0 & \frac{2}{3} \end{bmatrix},$$

then $A = LU$. Since their product equals A, the matrices L and U form a decomposition of A.

9.1 A Singularly Valuable Matrix Decomposition

The integer 315 can be factored as—$(3^2)(5)(7)$ or $(9)(5)(7)$. In a similar way, there can be more than one way to decompose a matrix. In this section, we are interested in a particular decomposition called the *singular value decomposition* (SVD), which has been a significant mathematical tool in both practical and theoretical settings. We'll assume we have a matrix calculator that has an SVD button, which can produce the SVD of a matrix. Feel free to read more about how to compute the SVD in a book or on the Internet, if you are interested.

What's returned by the matrix calculator after the SVD button is pressed? The SVD is defined by three matrices and can be found for any matrix. Let's consider a matrix that has more rows than columns:

$$B = \begin{bmatrix} 1 & 2 & 3 \\ 4 & 5 & 6 \\ 7 & 8 & 9 \\ 10 & 11 & 0 \end{bmatrix}.$$

Performing the SVD on a matrix returns three matrices, which are denoted U, Σ, and V. For B, our SVD calculator would calculate

$$U = \begin{bmatrix} -0.1520 & 0.2369 & 0.8684 \\ -0.3940 & 0.3626 & 0.2160 \\ -0.6359 & 0.4883 & -0.4365 \\ -0.6459 & -0.7576 & 0.0936 \end{bmatrix},$$

$$\Sigma = \begin{bmatrix} 21.0244 & 0 & 0 \\ 0 & 7.9811 & 0 \\ 0 & 0 & 0.5255 \end{bmatrix}, \quad \text{and}$$

$$V = \begin{bmatrix} -0.6012 & -0.6881 & -0.4064 \\ -0.3096 & -0.2682 & 0.9122 \\ -0.7367 & 0.6742 & -0.0518 \end{bmatrix}.$$

The product $U \Sigma V$ will equal B, where U is a matrix with the same number of rows and columns as B and V and Σ will have the same number of columns as U. Further, V and Σ have the same number of rows as columns. Σ is a *diagonal matrix*, which means it only contains nonzero values on the diagonal, which are called the *singular values*. We will assume that the singular values of Σ are ordered from highest to lowest with the highest appearing in the upper lefthand corner of the matrix. It is interesting that the columns of U and the rows of V^T in the SVD are related to our computations for PCA in Section 8.4. While the matrix product $U \Sigma V$ equals B, the three matrices can be used to approximate B, but that's getting a bit ahead of ourselves.

Before discussing the approximation of matrices, let us think a bit about the approximation of numbers. What do we mean when we say a number is a good approximation to another number? Generally, we are referring to distance. For example, 3.1111 would be considered a closer approximation to π than 3.0.

Similarly, we can define distance for matrices. One such measure is the Frobenius norm, which is denoted as $\|A\|_F$, defined as the square root of the sum of the squares of all the elements of the matrix. This is the distance from the matrix A to the zero matrix and can be written as

$$\|A\|_F = \sqrt{\sum_{i=1}^{m}\sum_{j=1}^{n}|a_{ij}|^2},$$

where A is an $m \times n$ matrix. Let's look at an example. If

$$A = \begin{bmatrix} 2 & 1 \\ 2 & -4 \end{bmatrix},$$

then

$$\|A\|_F = \left(2^2 + 1^2 + 2^2 + (-4)^2\right)^{1/2} = 5.$$

We square every element of the matrix, sum them, and take the square root. We have a norm, which we defined for vectors in Chapter 4. Matrix norms have similar properties.

With the Frobenius norm at our disposal, we are about to see the power of the SVD. First, we need to discuss what it means for a matrix to have rank k. A matrix has rank 1 if every row is a multiple of the first row. The matrix

$$M = \begin{bmatrix} 1 & 1 & 1 \\ 0.5 & 0.5 & 0.5 \\ 2 & 2 & 2 \end{bmatrix}$$

is an example of a rank 1 matrix.

We can also talk about a rank k matrix. A matrix has rank k if there exist k rows such that all remaining rows are a linear combination of those k rows, and k is the smallest such number. For instance,

$$M = \begin{bmatrix} 1 & 1 & 1 \\ 1 & 2 & 1 \\ 2 & 3 & 2 \end{bmatrix}$$

is an example of a rank 2 matrix. This is true since you can add the first and second rows to form the third, and no row is a constant multiple of another row.

The SVD gives the closest rank 1 approximation of a matrix as measured by the Frobenius norm. Finding this rank 1 matrix with the SVD is easy. We take the first column of U, denoted $\mathbf{u}$, the largest singular value (diagonal element) of Σ, denoted σ_1 and the first row of V, denoted $\mathbf{v}$.

Referring to the matrices U, Σ, and V found for the matrix B considered earlier in this section,

$$\mathbf{u} = \begin{bmatrix} -0.1520 \\ -0.3940 \\ -0.6359 \\ -0.6459 \end{bmatrix}, \sigma_1 = 21.0244, \text{ and}$$

$$\mathbf{v} = \begin{bmatrix} -0.6012 & -0.6881 & -0.4064 \end{bmatrix}.$$

To form the closest rank 1 approximation under the Frobenius norm, we compute $B_1 = \sigma_1 \mathbf{u} \mathbf{v}$:

$$B_1 = 21.0244 \begin{bmatrix} -0.1520 \\ -0.3940 \\ -0.6359 \\ -0.6459 \end{bmatrix} \begin{bmatrix} -0.6012 & -0.6881 & -0.4064 \end{bmatrix}$$

$$= \begin{bmatrix} 1.9216 & 2.1995 & 1.2989 \\ 4.9796 & 5.6997 & 3.3660 \\ 8.0376 & 9.2000 & 5.4330 \\ 8.1641 & 9.3449 & 5.5186 \end{bmatrix}.$$

We can verify that this is a rank 1 matrix by noting that the first row of B_1 is $\mathbf{v}$ times the product of the largest singular value (which equals 21.0244) and the first element of $\mathbf{u}$ (which equals -0.1520). Similarly, the second row of B_1 is $\mathbf{v}$ times the product of 7.9811 and the second element of $\mathbf{u}$ (which equals -0.3940). So, each row of B_1 is a multiple of $\mathbf{v}$ and is rank 1. It is a fact that a matrix produced by multiplying a column vector by a row vector will be a rank 1 matrix. Can you see why this would be true?

The SVD gives us much more than the best rank 1 approximation. It can also give us the best rank k approximation for any k, as long as k is less than or equal to the rank of our matrix A. How do you form it? You'll produce three matrices. The first matrix U_k is the first k columns of U. The second matrix Σ_k is a submatrix formed from the first k rows and k columns of Σ, which forms a $k \times k$ matrix. Then, V_k equals the first k rows of V. Our rank k approximation is $U_k \Sigma_k V_k$.

Let's practice again on the SVD of the matrix B. The rank of B is 3, so we an ask for, at most, a rank 3 matrix, which would equal B. Let's produce a rank 2 matrix $B_2 = U_2 \Sigma_2 V_2$. Then

$$U_2 = \begin{bmatrix} -0.1520 & 0.2369 \\ -0.3940 & 0.3626 \\ -0.6359 & 0.4883 \\ -0.6459 & -0.7576 \end{bmatrix},$$

$$\Sigma_2 = \begin{bmatrix} 21.0244 & 0 \\ 0 & 7.9811 \end{bmatrix}$$

and $V_2 = \begin{bmatrix} -0.6012 & -0.6881 & -0.4064 \\ -0.3096 & -0.2682 & 0.9122 \end{bmatrix}.$

So,

$$B_2 = \begin{bmatrix} -0.1520 & 0.2369 \\ -0.3940 & 0.3626 \\ -0.6359 & 0.4883 \\ -0.6459 & -0.7576 \end{bmatrix} \begin{bmatrix} 21.0244 & 0 \\ 0 & 7.9811 \end{bmatrix}$$

$$\times \begin{bmatrix} -0.6012 & -0.6881 & -0.4064 \\ -0.3096 & -0.2682 & 0.9122 \end{bmatrix}$$

$$= \begin{bmatrix} 1.3362 & 1.6923 & 3.0236 \\ 4.0836 & 4.9235 & 6.0059 \\ 6.8310 & 8.1546 & 8.9881 \\ 10.0362 & 10.9668 & 0.0025 \end{bmatrix}.$$

Let's return to the equation $B = U \Sigma V$ and express it in an alternate form:

$$B = 21.0244 \begin{bmatrix} -0.1520 \\ -0.3940 \\ -0.6359 \\ -0.6459 \end{bmatrix} \begin{bmatrix} -0.6012 & -0.6881 & -0.4064 \end{bmatrix}$$

$$+ 7.9811 \begin{bmatrix} 0.2369 \\ 0.3626 \\ 0.4883 \\ -0.7576 \end{bmatrix} \begin{bmatrix} -0.3096 & -0.2682 & 0.9122 \end{bmatrix}$$

$$+ 0.5255 \begin{bmatrix} 0.8684 \\ 0.2160 \\ -0.4365 \\ 0.0936 \end{bmatrix} \begin{bmatrix} -0.7367 & 0.6742 & -0.0518 \end{bmatrix}.$$

Multiplying a column vector by a row vector results in a rank 1 matrix. The sum we just used expresses B as the sum of rank 1 matrices. The coefficients of the terms are the singular values and the third rank 1 matrix is multiplied only by 0.5255. Maybe we can create a good approximation of B by dropping this third term in the sum, which would produce B_2 that we computed earlier.

This idea can be used to find the closest rank k approximation to a matrix. Suppose B is an 86×56 matrix with rank 56. If we want to produce the closest rank 48 matrix, then we form 48 rank 1 matrices. The first matrix equals the largest singular value times the rank 1 matrix formed by multiplying the first column of U with the first row of V. The second matrix is formed by multiplying the second largest singular value times the rank 1 matrix, which equals the product of the second column of U and the second row of V. We continue in this way until we form the 48th matrix by multiplying the 48th largest singular value by the rank 1 matrix formed by multiplying the 48th column of U with the 48th row of V. The sum of these 48 matrices is the closest rank 48 matrix.

How *close* is close under the Frobenius norm? Rather than looking at the numerical entries of a matrix, let's visualize the difference and use data related to images.

9.2 Feeling Compressed

Using the SVD on matrices that store grayscale data of an image is an application of this decomposition. If

$$A = \begin{bmatrix} 78 & 92 & 59 & 26 \\ 166 & 206 & 26 & 48 \\ 74 & 138 & 165 & 3 \\ 66 & 151 & 250 & 214 \end{bmatrix}, \tag{9.1}$$

then the corresponding image is depicted in Figure 9.1 (a). Each element of the matrix holds the grayscale value of the corresponding pixel in the image. So, the pixel in the upper lefthand corner of the picture has a grayscale value of 78.

Now we will consider a matrix A that is 648×509. The elements of the matrix correspond to the grayscale pixels of the Albrecht Dürer print seen in Figure 9.1 (b).

Now let's see what the matrix approximations look like as produced by the SVD. The matrix A_1, the closest rank 1 approximation to A, is visualized

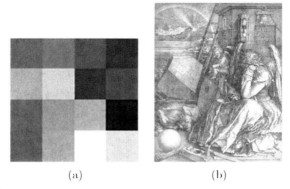

(a)   (b)

Figure 9.1. The grayscale image (a) of the data contained in the matrix A in (9.1). A 648 by 509 pixel image (b) of Albrecht Dürer's print *Melancholia*.

in Figure 9.2 (a). An advantage of this low rank approximation can be in the storage savings. The original matrix stored 648(509) = 329,832 numbers whereas A_1 can be stored with 648 numbers from U (the first column), 1 singular value, and 509 numbers from V (the first row). So, A_1 requires storing only 1,158 numbers or 0.35% of the original data. That's a very small fraction of the original data! Great savings on storage, but not a great representation of the original data.

Let's increase the rank of our approximation. The closest rank 10 approximation to A is depicted in Figure 9.2 (b). Let's look at the storage savings again. This matrix requires storing 10 column vectors of U, 10 singular values, and 10 row vectors of V. This correlates to a storage requirement of 6,480 + 10 + 5,090 numbers or about 3.5% of the original

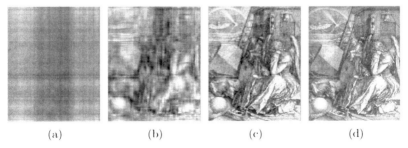

(a)   (b)   (c)   (d)

Figure 9.2. The closest rank 1 (a) and rank 10 (b) approximations to the Dürer print. In (c) and (d), we see the closest rank 50 approximation and the original print. Can you tell which is which?

image's storage requirements. The rank 10 image is a better approximation than the rank 1 approximation, as it should be. Still, it could be characterized as a computationally bleary–eyed image!

Let's crank up the rank of our approximation and jump to a rank 50 matrix. In Figure 9.2, we see the closest rank 50 approximation to A and the original print. Can you tell which is which? It would be even harder if we raised the rank of the approximation. This probably helps you sense that the closest rank 50 matrix is doing pretty well at approximating a rank 509 matrix.

The SVD typically is not the method used in compression algorithms. Other methods exist that can do an even better job. Still, this gives us a sense of how image compression can be done and why it sometimes can lead to a fuzzier picture than the original. In some cases, it is very difficult to visually see a difference between the original and the approximation. If we are sending pictures from the Mars Rover, then using a compressed image that looks essentially the same as the original may be quite helpful.

9.3 In a Blur

Data compression started with a matrix that stored a higher resolution image. Lower rank approximations to the matrix resulted in degradation in the data. When the data in the original matrix has noise, lower rank approximations can actually improve the quality of the data. Noise often occurs in data and measurement. Reducing it can help identify the underlying phenomenon.

We'll create our own data that contains noise. To begin, we'll generate exact data with the function

$$z = \sin\left[\frac{xy}{3}\right],$$

as adapted from lecture notes of Andrew Schultz of Wellesley College [15]. To construct the matrix, we take the z-values of this function over the grid formed between 0 and $3\pi/2$, in increments of $\pi/30$ in the both the x- and y-directions. We can then linearly interpolate between the points. This produces the graph seen in Figure 9.3 (a). To introduce noise, we'll add random values between -0.1 and 0.1 to every element of the matrix. This produces the image in Figure 9.3 (b).

As before, we'll construct a rank k approximation to the matrix. But what k should we choose? To aid in this decision, we turn to a graph of the

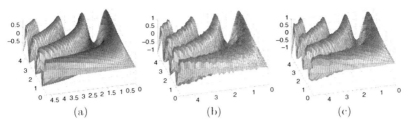

Figure 9.3. A 3D graph (a), with noise (b), that is reduced (c) using the SVD.

singular values seen in Figure 9.4. Looking at the plot, we see a steep drop-off in values beginning at the seventh largest singular value. That drop-off is the signal of a k to choose.

We'll take $k = 8$ and construct the rank 8 matrix approximation with the SVD. Now, we plot the rank 8 matrix approximation as seen in Figure 9.3 (c). The noise is reduced, though we do not entirely regain the image in Figure 9.3 (a). This technique is used when you may know there is noise but don't know where or how much. So, having improvement in the data like that seen in Figures 9.3 (c) and (b) can be important. The blurring of the image that occurred in compression aids in reducing noise.

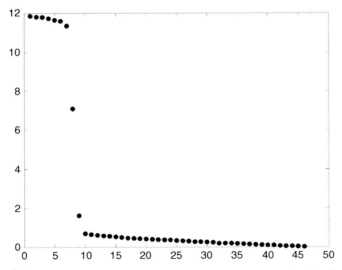

Figure 9.4. Singular values of the matrix containing the data graphed in Figure 9.3 (b).

(a) (b)

Figure 9.5. A color image of a mandrill (a), and then, in (b), after downsampling using the SVD for clustering. Image by Malene Thyssen (commons.wikimedia.org/wiki/File:Mandril.jpg) CC BY 2.5 (creativecommons.org/licenses/by/2.5/deed.en).

9.4 Losing Some Memory

In Chapter 8, we saw how eigenvectors can be used to cluster. The SVD can also be used for clustering. An application of clustering is in downsampling images, which can reduce the memory necessary to save a picture. The image of the mandrill in Figure 9.5 (a) contains pixels of various shadings. We will create a picture with pixels containing only the colors of white and black. Which pixels become white and which become black? Clustering will decide.

There is more than one way to store a color image. We'll use three matrices to store the red, green, and blue channels of a color image. So, one pixel will be stored as a triplet like (15, 255, 10), which, in this case, is predominantly green with some red and a little less blue. To store the pixel information of an $n \times m$ image, we'll form an $(nm) \times 3$ matrix T where each row contains the red, green, and blue intensities of a pixel. Since the elements of T are all nonnegative, we recenter them prior to clustering. Since the values range between 0 and 255, we will subtract 127 from each value.

Now, we are ready for the SVD, which can be found for any matrix. This flexibility in the SVD is important since we have a very long and skinny matrix. Now, we form the SVD and get $U \Sigma V$. The dimensions of U and T

are the same. We will take the first column of U, which is also called a left singular vector. Its first element can be associated with the same pixel that is associated with the first element in T. We can associate the second element of the singular vector with the same pixel as is associated with the second element of T. Associating every pixel in this way, we are ready to cluster. Pixels that correspond to negative elements in the singular vector are put in one cluster. All other pixels are put in the other cluster. Then we decide on a color for each pixel in each cluster. We see this in Figure 9.5 (b) where the colors are chosen to be white and black. The color could also be chosen to be the average color of the pixels in the cluster.

In this chapter, we decomposed a matrix into factors that give insight to the matrix. This allows us to reduce the amount of storage necessary to save a matrix. In another context, the quality of the information was improved. By the end of the next chapter, we will see how Google uses matrices to mine information from its massive database, the World Wide Web. Then, in the remaining chapters, we look at applications of data analytics.

10

What Are the Chances?

The world is infused with chance. Random events have underlying probabilities. A fair flip of a coin has the same chance of landing heads as tails. In this chapter, we use linear algebra to analyze situations that involve randomness.

10.1 Down the Chute

Ever play Monopoly for several days? How often does this happen? How short a game is possible? These types of questions can be analyzed using matrices to determine such things as average and shortest game length. Let's try this with a few different games.

First, we will look at Chutes and Ladders, also known as Snakes and Ladders. Chutes and Ladders originated in India as a game of knowledge and was known as Jñána Chaupár. Landing on a virtue resulted in climbing a ladder toward the god Vishnu. Rather than sliding down a chute, the game had a player swallowed by a snake, which resulted in death and a new start. This game entered middle-class Victorian parlors of England in the nineteenth century as Snakes and Ladders. Board game pioneer Milton Bradley introduced the game as Chutes and Ladders in the United States in 1943 promoting it as the "improved new version of snakes and ladders, England's famous indoor sport."

The game is played with two or more players on a board with numbered squares on a grid as seen in Figure 10.1. Players begin on the starting square and the first to reach the finish square wins. This journey from beginning to end is helped and hindered by ladders and chutes (or snakes) that appear on the board.

Let's see how linear algebra can aid us in analyzing the game. We'll look at a smaller game board to simplify the computations. We'll analyze the

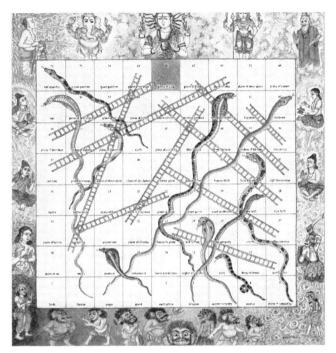

Figure 10.1. Game of Snakes and Ladders on cloth from the 19th century in India. From "Hinduism Endures: 1100 to 1850," 2009, Hinduism Today, www.hinduismtoday.com. Copyright 2009 by Himalayan Academy. Reprinted with Permission.

game board in Figure 10.2. The game starts on square 1. Reaching square 9 results in winning. On each turn, roll a die. If you roll 1 or 2, do not move and stay on your current square. If you roll 3 or 4, move ahead one square. If you roll 5 or 6, move ahead two squares. If you land on square 4, you slide down the snake to square 2. If you land on square 6, you climb the ladder to square 8. When are you are on square 8, you can roll 3, 4, 5, or 6 and land at square 9 to complete the game.

We will ask a couple of questions, which will be answered with linear algebra. What is the minimum number of moves in the game? How many moves are necessary for it to be 50% likely that someone has won? To answer these, we'll use *Markov chains*, which use matrices in which the elements correspond to probabilities. This means that the sum of each row of every matrix will be 1.

Our first step is forming a *transition matrix*, which contains the probabilities of going from one square to another. Row *i* of the matrix corresponds

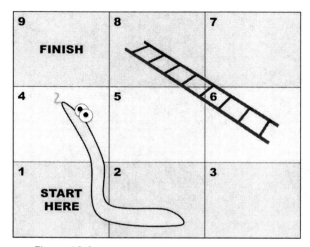

Figure 10.2. A small game of Snakes and Ladders.

to being on square i in the game. In the jth column of row i, we place the probability of going to square j from square i when the die is rolled.

Let's form the first row of the matrix. This corresponds to being on square 1. So, we have a 1/3 chance of staying on square 1 and a 1/3 chance of moving to square 2 or square 3. So, the first row is

$$\begin{bmatrix} 1/3 & 1/3 & 1/3 & 0 & 0 & 0 & 0 & 0 & 0 \end{bmatrix}.$$

What happens on square 2? There is a 1/3 chance of staying on square 2 or moving to square 3. There is also a 1/3 probability of landing on square 4, which actually returns us to square 2. So, the second row is:

$$\begin{bmatrix} 0 & 2/3 & 1/3 & 0 & 0 & 0 & 0 & 0 & 0 \end{bmatrix}.$$

Verify that the transition matrix A is

$$A = \begin{bmatrix} 1/3 & 1/3 & 1/3 & 0 & 0 & 0 & 0 & 0 & 0 \\ 0 & 2/3 & 1/3 & 0 & 0 & 0 & 0 & 0 & 0 \\ 0 & 1/3 & 1/3 & 0 & 1/3 & 0 & 0 & 0 & 0 \\ 0 & 0 & 0 & 0 & 0 & 0 & 0 & 0 & 0 \\ 0 & 0 & 0 & 0 & 1/3 & 0 & 1/3 & 1/3 & 0 \\ 0 & 0 & 0 & 0 & 0 & 0 & 0 & 0 & 0 \\ 0 & 0 & 0 & 0 & 0 & 0 & 1/3 & 1/3 & 1/3 \\ 0 & 0 & 0 & 0 & 0 & 0 & 0 & 1/3 & 2/3 \\ 0 & 0 & 0 & 0 & 0 & 0 & 0 & 0 & 1 \end{bmatrix}.$$

To analyze the game, we begin with a vector, containing probabilities that sum to 1, representing where the game begins, which is (with probability 1) on square 1. This corresponds to the vector

$$\mathbf{v} = \begin{bmatrix} 1 & 0 & 0 & 0 & 0 & 0 & 0 & 0 & 0 \end{bmatrix}.$$

Where will we be after one turn? This involves computing $\mathbf{v}_1 = \mathbf{v}A$, so

$$\mathbf{v}_1 = \begin{bmatrix} 1/3 & 1/3 & 1/3 & 0 & 0 & 0 & 0 & 0 & 0 \end{bmatrix}.$$

This indicates that you have a 1/3 chance of being at square 1, 2, or 3 after 1 move. Now, $\mathbf{v}_2 = \mathbf{v}_1 A$, so

$$\mathbf{v}_2 = \begin{bmatrix} 0.1111 & 0.4444 & 0.3333 & 0 & 0.1111 & 0 & 0 & 0 & 0 \end{bmatrix}.$$

This indicates that after two moves, there is about an 11% chance you are still at square 1, a 44% chance you of being at square 2, a 33% chance of being at square 3, and an 11% chance of being at square 5.

Let's find the minimum number of moves in the game. To find this, we compute $\mathbf{v}_n = A\mathbf{v}_{n-1}$ until the last element in $\mathbf{v}_n$ is nonzero. This happens when $n = 4$. Look at the game and verify that one could win it in four moves. What's the probability of this happening? Just under 4% since

$$\mathbf{v}_4 = \begin{bmatrix} 0.0123 & 0.4074 & 0.2593 & 0 & 0.1481 & 0 & 0.0617 & 0.0741 & 0.0370 \end{bmatrix}.$$

Our other question was how many moves are necessary for it to be 50% likely that someone has won? We compute $\mathbf{v}_n = A\mathbf{v}_{n-1}$ until the last element in $\mathbf{v}_n$ first becomes larger than 0.50. This happens for

$$\mathbf{v}_{10} = \begin{bmatrix} 0.0000 & 0.1854 & 0.1146 & 0 & 0.0708 & 0 & 0.0436 & 0.0700 & 0.5156 \end{bmatrix}.$$

So, it is possible that you could win in four moves but generally it will take ten. If you know how long a turn takes, you can even compute about how long the game will often last!

Suppose, we decide to change the game. Rolling 1 or 2 still results in not moving. Now, rolling 3, 4, or 5 will move your piece one square. Rolling a 6 moves you two squares. The minimum length of the game is still four moves but that now will only occur about 1% of the time. How about game length? Now, after 15 moves you are 50% likely to be done the game.

Let's answer these questions for a larger game board as seen in Figure 10.3. If we set up the transition matrix, we can analyze the game. We'll assume you again have a 1/3 chance of staying on your current square, moving 1, or moving 2 squares forward. Now, the minimum length of the game is six turns happening with about a 1% chance.

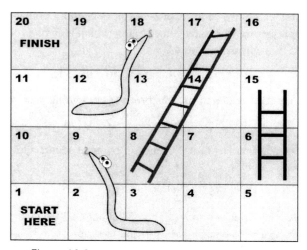

Figure 10.3. A larger game of Snakes and Ladders.

Something about the board becomes evident when you look at $\mathbf{v}_n$ for increasing n. Regardless of the value of n, the 10th and 11th elements of $\mathbf{v}_n$ equal 0. This means you never have a chance to reach these squares. Take a moment and look at the board. It was constructed for this to happen. One could also find squares that are more and less likely to be reached. This could help you determine rewards to offer on different squares. In the end, Markov chains can tell you a lot about a game—even helping you improve or design it.

Break out your board games, see which can be modeled with a Markov chain, and answer your own questions and maybe add your own rules!

10.2 Google's Rankings of Web Pages

Search engines like Google's are big business. In 2005, Google founders Larry Page and Sergey Brin each had a net worth of more than 10 billion dollars. Less than 10 years earlier in 1998, the duo had dropped out of graduate school at Stanford and were working out of a garage on their search engine business.

There is also big business for businesses in search engines. How often do you search the Internet when shopping or researching items to purchase? The *Wall Street Journal* reported that 39% of web users look at only the first search result returned from a search engine and 29% view only a few results.

If a business owner finds the company coming up on the second page of results, there is understandable concern. Understanding the algorithm behind many search engines can help explain why one page is listed before another.

What do we want to have returned from a query to a search engine? We need web pages to be relevant to our query. We also need a sense of the quality of a web page, and this is where we will focus our attention. With billions of web pages out there, how can we possibly determine the quality of a page?

Google tackles this issue with the use of the PageRank algorithm, developed by Page and Brin. Google determines the popularity of a web page by modeling Internet activity. If you visit web pages according to Google's model, which pages would you end up at the most? The frequency of time spent on a web page yields that page's PageRank.

What is Google's model of surfing? Is someone tracking your surfing to build the model? Google models everyone as being a random surfer by assuming that you randomly choose links to follow. In this way, the model stays away from making decisions based on preferred content.

The PageRank model assumes that you have an 85% chance of following a hyperlink on a page, and a 15% chance of jumping to any web page in the network (with uniform probability), any time you are on a web page with links on it. If you are on a web page with no links on it, like many PDF files or even video or audio files, then you are equally likely to jump anywhere on the Internet. Such a page is called a dangling node. You could even jump back to your current page! That can seem odd. Why allow someone to jump back to the same page? Why would you ever do that?

The better question is why let the model do this. Why? To guarantee an answer. This means no matter how the World Wide web is organized, we always have a ranking and are never unable to know how to rank the web pages. What guarantees this is a theorem based on the linear algebra Google uses for the PageRank algorithm.

Let's see this model of surfing in action on the network in Figure 10.4. The graph represents six web pages, with the vertices or nodes of the graph representing web pages. A directed edge from one vertex to another represents a link from one web page to another. So, on web page 1 there is a link that you can click and go to web page 4. On web page 4, there are links to web pages 2 and 5. Web page 6 has no links.

By the PageRank model, if you are on web page 6, then you will jump, with equal probability, to any web page in the network. Said another way,

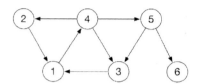

Figure 10.4. A small network of web pages.

you have a 1/6 probability (since there are six pages in this web network) of jumping to any given web page. If you are on web page 1, you have an .85 probability of clicking the link to take you to web page 4 and a .15/6 probability that you jump to any web page in the network.

This type of thinking gives us the probability of going from any web page to any other web page. This allows us to use Markov chains. We'll create a matrix G in which entry g_{ij} equals the probability of being at web page i and moving to web page j.

Let's compute the first row of G for the network in Figure 10.4. Entries in this row give the probabilities of jumping from web page 1 to page i. The entry in the first column, g_{11}, equals the probability of moving from web page 1 to web page 1. This only happens through *teleportation*, which is the term Page and Brin used for jumping from one web page to another without following a link. In our model of surfing this network, 15% of the time we teleport, and at that time every page is an equally likely destination. So, each page has a $(15/100)(1/6) = 1/40$ chance of being visited through teleportation from web page 1. Hyperlinks on a page are clicked 85% of the time. Web page 1's only link is to web page 4. So, web page 4 will be visited through teleportation $1/40$ of the time and through web page 1's hyperlink 85% of the time. So, there is a $35/40$ chance of visiting web page 4 after your first step of surfing. Therefore, the first row of G is

$$\mathbf{g}_1 = \begin{bmatrix} 1/40 & 1/40 & 1/40 & 35/40 & 1/40 & 1/40 \end{bmatrix}.$$

Before producing the entire matrix, let's produce the sixth row which assumes you are at the dangling node. So, we must teleport. Since all web pages are equally likely as a destination, each web page has a 1/6 probability of being visited from web page 6. So, the sixth row of G is

$$\mathbf{g}_6 = \begin{bmatrix} 6/36 & 6/36 & 6/36 & 6/36 & 6/36 & 6/36 \end{bmatrix}.$$

Similar logic for all the rows produces

$$G = \begin{bmatrix} 1/40 & 1/40 & 1/40 & 35/40 & 1/40 & 1/40 \\ 35/40 & 1/40 & 1/40 & 1/40 & 1/40 & 1/40 \\ 35/40 & 1/40 & 1/40 & 1/40 & 1/40 & 1/40 \\ 1/40 & 37/120 & 37/120 & 1/40 & 37/120 & 1/40 \\ 1/40 & 1/40 & 18/40 & 1/40 & 1/40 & 18/40 \\ 6/36 & 6/36 & 6/36 & 6/36 & 6/36 & 6/36 \end{bmatrix},$$

which is called a *Markov transition matrix*.

Let's assume we start at web page 1. We represent this with the vector

$$\mathbf{v} = \begin{bmatrix} 1 & 0 & 0 & 0 & 0 & 0 \end{bmatrix},$$

which means that there is a probability of 1 that we are at web page 1. To determine the probability of visiting each web page in the network after one step, compute

$$\mathbf{v}G = \mathbf{v}_1 = \begin{bmatrix} 1/40 & 1/40 & 1/40 & 35/40 & 1/40 & 1/40 \end{bmatrix}.$$

The probabilities of visiting the web pages after an additional step through the network are

$$\mathbf{v}_1 G = \mathbf{v}_2 = \begin{bmatrix} 0.0710 & 0.2765 & 0.2871 & 0.0498 & 0.2765 & 0.0392 \end{bmatrix}.$$

Note that

$$\mathbf{v}_2 = \mathbf{v}_1 G = \mathbf{v}G^2.$$

So, $\mathbf{v}_{100} = \mathbf{v}G^{100}$. For this example, $\mathbf{v}_{50} = \begin{bmatrix} 0.2677 & 0.1119 & 0.1595 & 0.2645 & 0.1119 & 0.0845 \end{bmatrix}$, which indicates that after 50 steps, we expect to be at web page 1 about 26.8% of the time. Interestingly, for $n > 50$, the vector $\mathbf{v}_n$ will equal $\mathbf{v}_{50}$ when truncated to five decimal places.

PageRank defines the rating of a web page (the quality of it) as the probability assigned to that web page through this type of process. Said another way, if we keep iterating, what does a web page's probability settle to? This is its rating. The higher the rating the higher the quality of the page.

The vector we are finding is called the *steady-state vector*. The steady-state vector has the property that $\mathbf{v} \approx \mathbf{v}G$. It is a left eigenvector of G since we are multiplying the vector on the left. The associated eigenvalue is 1 since the vector is not changing when multiplied by G.

How does Google know for any network of web pages it can find such a steady-state vector? The PageRank model assumes a surfer can, at any stage, jump to any other web page. This results in all positive entries in the transition matrix, which according to Perron's theorem guarantees a unique steady-state vector.

> **Perron's theorem** *Every square matrix with positive entries has a unique eigenvector with all positive entries; this eigenvector's corresponding eigenvalue has only one associated eigenvector, and the eigenvalue is the largest of the eigenvalues.*

Let's walk through the use of this theorem. First, the Google matrix G is a square matrix and all of its entries are real numbers. The last requirement for this theorem to be used is that the matrix must have all positive entries. Teleportation guarantees that this will always be true for any G corresponding to any network.

So, Perron's theorem applies to any Google matrix G. What does it tell us? There is a unique eigenvector with all positive entries that sum to 1. That is, one and only one left (or right) eigenvector of G has all positive entries and the eigenvalue associated with this eigenvector is the largest eigenvalue; the eigenvalue also only has one associated eigenvector. Let's see how this helps us.

It's easy to find an eigenvector of G with all positive entries. The rows of G sum to 1. Therefore, $G\mathbf{u} = \mathbf{u}$, where $\mathbf{u}$ is the column vector of all ones. That is, $\mathbf{u}$ is a right eigenvector of G associated with the eigenvalue 1.

Perron's theorem ensures that $\mathbf{u}$ is a right eigenvector with all positive entries, and hence its eigenvalue must be the largest eigenvalue. This means that there must also be a unique unit vector $\mathbf{v}$ that satisfies $\mathbf{v}G = \mathbf{v}$. It has all positive entries and will be unique. For a Google matrix G, the PageRank vector $\mathbf{v}$ is unique, and its entries can be viewed as probabilities for our desired ranking.

This guarantees that no matter how much the web changes or what set of web pages Google indexes, the PageRank vector can be found and will be unique.

Let's close by finding the PageRank vector for the network in Figure 10.5. Computing $\mathbf{v}_n$ for large n helps us find $\mathbf{v}$ such that $\mathbf{v}G = \mathbf{v}$ and yields

$$\mathbf{v} = \begin{bmatrix} 0.2959 & 0.1098 & 0.2031 & 0.2815 & 0.1098 \end{bmatrix}.$$

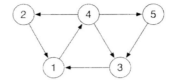

Figure 10.5. Another network of web pages.

So, a random surfer will visit page 1 approximately 30% of the time and page 2 about 11% of the time.

Pages 1 and 3 have the same indegree and outdegree, yet page 1 is linked from pages that, in the end, have higher PageRank. Page 4 receives a high PageRank because it is linked from page 1. If a surfer lands on page 1 (which occurs about 30% of the time), then 85% of the time the surfer will follow a link. Page 1 links only to page 4. The high PageRank of page 1 boosts the probability that page 4 is visited.

It pays to have a recommendation from a web page with high PageRank. There are companies that you can pay to boost your PageRank. This is accomplished in a variety of ways. You could try it yourself. Just figure out how to get a web page with high PageRank to link to your page. Get in the news, have CNN.com link to your page and who knows where you will land in the web results of Google. Then again, be careful what you do to get in the news or you may be viewing those web results while spending your time in a small cell with no windows!

10.3 Enjoying the Chaos

In the previous sections, we used matrices to analyze events with an element of randomness. Now, let's use randomness to produce mathematical art, beginning with the three squares below and stepping through the rules that follow.

1

          3

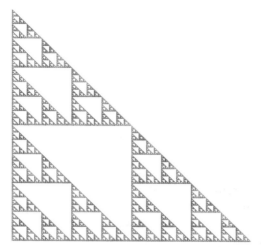

Figure 10.6. Sierpiński's triangle, a fractal, named after its founder, Waclaw Sierpiński.

Rules

1. Place a dot halfway between square 1 and 2.
2. Roll a die and place a new dot halfway between where you placed the last dot and square 1 if you roll 1 or 2, square 2 if you roll 3 or 4, or square 3 if you roll 5 or 6.
3. Return to Step 2.

Play a few times! What shape emerges? If you played long and accurately enough, the emerging image is a fractal known as Sierpiński's triangle seen in Figure 10.6.

The image contains three copies of the larger image. There is one at the top and two along the bottom. Magnifying an object and seeing similarities to the whole is an important property of fractals. An object with self-similarity has the property of looking the same as or similar to itself under increasing magnification.

How do we create this shape using linear algebra? Let's look carefully at the rules. Let's represent the squares in the game by the vectors $\mathbf{p}_1 = [\,0\ 1\,]^T$, $\mathbf{p}_2 = [\,0\ 0\,]^T$, and $\mathbf{p}_3 = [\,1\ 0\,]^T$. The game we just played, sometimes called the chaos game, entails taking our current vector $\mathbf{v}$ and letting the new vector equal half the distance between the current $\mathbf{v}$ and vectors for squares 1, 2, or 3, which we choose randomly. If we let $\mathbf{v}_n$ denote

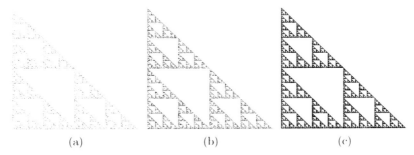

(a)   (b)   (c)

Figure 10.7. Sierpiński's triangle created with the chaos game visualized after 5,000 (a), 20,000 (b), and 80,000 (c) steps.

the new vector, then one step of our game is captured in

$$\mathbf{v}_n = (\mathbf{v} + \mathbf{p})/2,$$

where $\mathbf{p}$ is randomly chosen from $\mathbf{p}_1 = [0 \quad 1]^T$, $\mathbf{p}_2 = [0 \quad 0]^T$, and $\mathbf{p}_3 = [1 \quad 0]^T$.

Sierpiński's triangle forms after an infinite number of iterations but there comes a point after which a larger number of iterations no longer produces visible differences. Such changes are only perceived after zooming into regions. To see this visually, Figures 10.7 (a), (b), and (c) are this process after 5,000, 20,000, and 80,000 iterations.

We can also represent the chaos game as

$$T(\mathbf{v}) = \begin{bmatrix} 1/2 & 0 \\ 0 & 1/2 \end{bmatrix} (\mathbf{v} - \mathbf{p}) + \mathbf{p},$$

where $\mathbf{p}$ is randomly chosen to be $\mathbf{p}_1$, $\mathbf{p}_2$, or $\mathbf{p}_3$. $T(\mathbf{v})$ is a new point that we graph since it is part of the fractal. Then, we let $\mathbf{v}$ equal $T(\mathbf{v})$ and perform the transformation again. This looping produces the fractal.

This formula becomes more versatile when we add another term:

$$T(\mathbf{v}) = \begin{bmatrix} 1/2 & 0 \\ 0 & 1/2 \end{bmatrix} \begin{bmatrix} \cos\theta & -\sin\theta \\ \sin\theta & \cos\theta \end{bmatrix} (\mathbf{v} - \mathbf{p}) + \mathbf{p}.$$

Let's set θ to 5 degrees. This produces the image in Figure 10.8 (a). An interesting variation is to keep θ at 0 degrees when either points 2 or 3 are chosen in the chaos game. If point 1 is chosen, then we use θ equal to 5 degrees. This produces the image in Figure 10.8 (b).

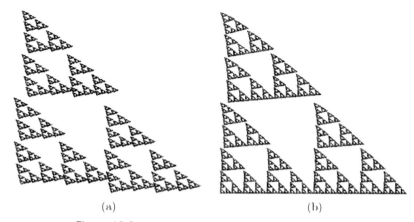

(a)   (b)

Figure 10.8. Sierpiński's triangle created with rotations.

Finally, let's use five points set as the corners of a regular pentagon. We'll also set $a = 2.65$ and $\theta = 0$. This produces the image in Figure 10.9 (a).

Want something more realistic? Then let's try a similar but different transformation. Now, we will have four transformations.

$$T_1(\mathbf{v}) = \begin{bmatrix} 0.85 & 0.04 \\ -0.04 & 0.85 \end{bmatrix} \mathbf{v} + \begin{bmatrix} 0 \\ 1.64 \end{bmatrix}$$

$$T_2(\mathbf{v}) = \begin{bmatrix} 0.20 & -0.26 \\ 0.23 & 0.22 \end{bmatrix} \mathbf{v} + \begin{bmatrix} 0 \\ 1.6 \end{bmatrix}$$

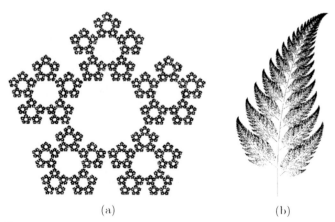

(a)   (b)

Figure 10.9. Variations on the chaos game make interesting fractals.

$$T_3(\mathbf{v}) = \begin{bmatrix} -0.15 & 0.28 \\ 0.26 & 0.24 \end{bmatrix} \mathbf{v} + \begin{bmatrix} -0.028 \\ 1.05 \end{bmatrix}$$

$$T_4(\mathbf{v}) = \begin{bmatrix} 0 & 0 \\ 0 & 0.16 \end{bmatrix} \mathbf{v}.$$

Given $\mathbf{v}$, we create the next point by choosing among T_1, T_2, T_3, and T_4 with probabilities 0.85, 0.07, 0.07 and 0.01, respectively. This creates the fractal known as Barnsley's fern seen in Figure 10.9 (b).

In this chapter, we've seen how randomness and a sense of chance can enable us to model aspects of the world via Markov chains and to even create beauty reminiscent of the natural world through fractals. In the next chapter, we look for information within random noise in the field of data mining.

11

Mining for Meaning

From smartphones to tablets to laptops and even to supercomputers, data is being collected and produced. With so many bits and bytes, data analytics and data mining play unprecedented roles in computing. Linear algebra is an important tool in this field. In this chapter, we touch on some tools in data mining that use linear algebra, many built on ideas presented earlier in the book.

Before we start, how much data is a lot of data? Let's look to Facebook. What were you doing 15 minute ago? In that time, the number of photos uploaded to Facebook is greater than the number of photographs stored in the New York Public Library photo archives. Think about the amount of data produced in the past two hours or since yesterday or last week. Even more impressive is how Facebook can organize the data so it can appear quickly into your news feed.

11.1 Slice and Dice

In Section 8.3, we looked at clustering and saw how to break data into two groups using an eigenvector. As we saw in that section, it can be helpful, and sometimes necessary for larger networks, to plot the adjacency matrix of a graph. In Figure 11.1, we see an example where a black square is placed where there is a nonzero entry in the matrix and a white square is placed otherwise.

The goal of clustering is to find maximally intraconnected components and minimally interconnected components. In a plot of a matrix, this results in darker square regions. We saw this for a network of about fifty Facebook friends in Figure 8.5 (b). Now, let's turn to an even larger network. We'll analyze the graph of approximately 500 of my friends on Facebook.

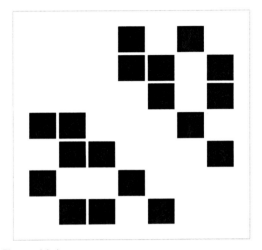

Figure 11.1. Plotting the adjacency matrix of a graph.

We see the adjacency matrix visualized in Figure 11.2 (a). Here we see little organization or a pattern of connectivity with my friends. If we partition the group into two clusters using the Fiedler method outlined in Section 8.3, after reordering the rows and columns so clusters appear in a group, we see the matrix in Figure 11.2 (b).

Remember, we partitioned the matrix into clusters. We see this with the smaller, but dense, matrix in the upper left and the larger matrix in the

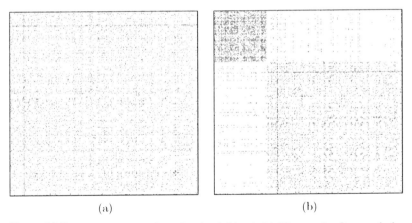

(a) (b)

Figure 11.2. The adjacency of 500 Facebook friends (a). The matrix after reordering the rows and columns into two clusters found using an eigenvector (b).

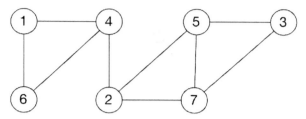

Figure 11.3. An undirected graph to cluster.

lower right. There are still connections outside each cluster. This is due to friendships that cross the groups. But, what are the groups? Clustering won't identify them. The smaller cluster is largely alumni from Davidson College (where I teach) and assorted faculty. The larger cluster contains friends from various other parts of my life—high school, college, graduate school and such. This isn't entirely true but overall the pattern is identifiable. So, we'll learn to break the matrix into more groups, using a method called *extended Fiedler*.

Let's return to the small network from Section 8.3, which is reprinted in Figure 11.3. As we saw earlier, we want to look at the eigenvector corresponding to the second smallest eigenvalue of the Laplacian matrix, which for this problem is

$$L = \begin{bmatrix} 2 & 0 & 0 & -1 & 0 & -1 & 0 \\ 0 & 3 & 0 & -1 & -1 & 0 & -1 \\ 0 & 0 & 2 & 0 & -1 & 0 & -1 \\ -1 & -1 & 0 & 3 & 0 & -1 & 0 \\ 0 & -1 & -1 & 0 & 3 & 0 & -1 \\ -1 & 0 & 0 & -1 & 0 & 2 & 0 \\ 0 & -1 & -1 & 0 & -1 & 0 & 3 \end{bmatrix}.$$

The eigenvector of interest is

$$\begin{bmatrix} 0.4801 \\ -0.1471 \\ -0.4244 \\ 0.3078 \\ -0.3482 \\ 0.4801 \\ -0.3482 \end{bmatrix}.$$

The rows with the same sign are placed in the same cluster. So our vector can be simplified to

$$\begin{bmatrix} + \\ - \\ - \\ + \\ - \\ + \\ - \end{bmatrix},$$

indicating the nodes 1, 4, and 6 cluster together. This leaves nodes 2, 3, 5, and 7 for the other cluster.

To break these two clusters into more clusters, we now look at the eigenvector associated with the next smallest eigenvalue, which is

$$\begin{bmatrix} -0.2143 \\ 0.6266 \\ -0.6515 \\ 0.2735 \\ 0.0900 \\ -0.2143 \\ 0.0900 \end{bmatrix}.$$

We again look at the sign of each element but now of the signs of the entries of the eigenvectors associated with the second and this third smallest eigenvalues. This gives

$$\begin{bmatrix} + & - \\ - & + \\ - & - \\ + & + \\ - & + \\ + & - \\ - & + \end{bmatrix}.$$

We now place rows with the same sign pattern in both columns in the same cluster. So, we cluster nodes 1 and 6 together, since they are both positive in the first column and negative in the second. We also produced a *singleton*, which is a cluster of only one node. Node 4 is a singleton, since it is the only row with positive elements in both columns. Nodes 2, 5, and 7 cluster

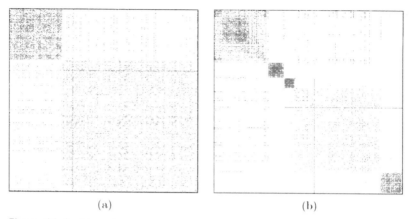

(a)   (b)

Figure 11.4. Clustering 500 Facebook friends using one eigenvector (a) and four eigenvectors (b).

together, since they have a negative in the first column and a positive in the second. This leaves node 3 as another singleton. We did not gain much by this clustering—only two singletons. The introduction of singletons is often a sign of creating too many clusters with a dataset. Clustering can be as much an art as a science and depends, in part, on the goal of the data mining.

While unnecessary for this problem, we can create more clusters. We use the eigenvectors associated with the second, third, . . . through the $(k + 1)$ st smallest eigenvalues and cluster nodes with rows that contain the same sign pattern as we just did with $k = 2$. Returning to our two eigenvectors, note that we attained four clusters but, it could happen that the sign patterns only yield three clusters. We are guaranteed up to 2^k clusters when we use k eigenvectors for the clustering.

The power of extended Fiedler is more apparent when we cluster networks much larger than that in Figure 11.3. Let's return to my Facebook friends. In Figure 11.4 (a), we repeat the clustering formed with one eigenvector. In Figure 11.4 (b), we see the clustering formed with four eigenvectors. If five vectors are used, several singletons are formed. With four eigenvectors, a total of 12 clusters, as opposed to the 16 that is the upper bound, are formed. The clusters form groups that are largely identifiable by their communities. Many other clustering algorithms exist, each with its own strength and type of community it tends to identify. So, dissatisfied with what clusters are formed? It may be that you need another algorithm.

11.2 Movie with not Much Dimension

Now, let's combine the ideas of Section 5.2 where we found similar movies with the compression power of Principal Component Analysis (PCA) discussed in Section 8.4. One of the huge benefits of PCA is its ability to capture important trends in data with large savings in storage. In Section 5.2, we compared movies in a space of dimension more than 900. The size of the space is connected to the number of users in the dataset. The MovieLens database consists of 100,000 ratings (1–5) from 943 users on 1682 movies.

Let's again find similar movies. The data will be stored in a matrix with 1,682 rows and 943 columns. So, each row corresponds to a movie and contains the ratings by every user for that movie. To use PCA, as discussed in Section 8.4, we first subtract the mean of each row from each element in that row. Each row contains 943 user ratings. Let's use PCA to create a lower dimensional approximation to the data.

Let's again consider the 1977 film *Star Wars*. This time, however, we use PCA to create a 50-dimensional approximation to the ratings vector. That is, we use 50 eigenvectors in PCA. When we do this, we find the three most similar films are still *Return of the Jedi* (1983), *The Empire Strikes Back* (1980), and *Raiders of the Lost Ark* (1981) using cosine similarity of the lower dimensional vectors. For *Snow White and the Seven Dwarfs*, the full vectors find the top three to be *Beauty and the Beast* (1991), *Cinderella* (1950), and *Pinocchio* (1940). When PCA is used, the ordering changes to *Cinderella* (1950), *Pinocchio* (1940), and then *Beauty and the Beast* (1991).

Is one better than the other? More testing and even validation, if possible, would be necessary. Lower dimensional approximations have the ability to remove noise but also tend to blur and obscure detail in the full matrix. So, there is a balance between reducing noise and losing too much detail. There is both an art and a science to using PCA. Yet, notice the amount of information we gleaned from a lower approximation, pointing to the power of such a method.

11.3 Presidential Library of Eigenfaces

Let's apply PCA in the context of image processing. As we have seen previously, images can be stored as vectors or matrices. When stored as a vector, the dimension is the number of pixels. PCA uses linear combinations of eigenvectors to recapture an approximation to the original faces.

Figure 11.5. Library of twelve Presidents of the United States.

The power is in the compression. We may need only 50 or 100 *eigenfaces*, as they are called, to create strong approximations to a set of 10,000 faces.

For us to see this, we will take a small dataset of portraits of presidents of the United States as seen in Figure 11.5. Here we see the connection of PCA with the singular vector decomposition (SVD). Each image has the same pixel resolution with n rows and m columns. Then, each picture is converted to a vector.

The average image is computed and the difference between every image and the average is also calculated. For example, the average president is seen in Figure 11.6 (a). In Figure 11.6 (b), we see the average image in (a) subtracted from the image of President Kennedy found in Figure 11.5.

Then each of these mean-subtracted vectors becomes a column of a matrix P. We then form $C = PP^T$, which is named C since this matrix is known as the covariance matrix. We are interested in the eigenvectors of C.

Because the number of rows in P equals the number of pixels in each image, which in this case is 50,000, P is a $50,000 \times 50,000$ matrix. Finding eigenvectors would be prohibitively time-consuming even for a supercomputer. However, when there are N images in the library, there will be at most $N - 1$ eigenvectors with non-zero eigenvalues. So, our presidential library will have at most eleven eigenvectors with non-zero eigenvalues.

To deal with the size of C, we take, instead, $P^T P$, which is only 12×12. Fortunately, the eigenvalues are the same for both systems. We need the eigenvectors, though. If we have an eigenvector $\mathbf{u}_i$ of $P^T P$, then

$$P^T P \mathbf{u}_i = \lambda_i \mathbf{u}_i.$$

(a)   (b)

Figure 11.6. The average image computed over the library of twelve U.S. Presidents seen in Figure 11.5 (a) and the average image subtracted from the image of President Kennedy (b).

Multiplying both sides by P, we get

$$P P^T P \mathbf{u}_i = P \lambda_i \mathbf{u}_i, \text{ so } P P^T (P \mathbf{u}_i) = \lambda_i (P \mathbf{u}_i),$$

which means if $\mathbf{u}_i$ is an eigenvector of $P^T P$ then $P \mathbf{u}_i$ is an eigenvector of $C = P P^T$. We went from looking for eigenvectors of a $50,000 \times 50,000$ matrix to a 12×12 matrix.

Figure 11.7 depicts six eigenfaces (which again are simply eigenvectors) of the six largest eigenvalues of C for our presidential library of images. Here we make the connection to eigenvalues of C and singular values. The singular values of P are the square roots of the nonzero eigenvalues of C. How this is helpful lies in what we learned about singular values in Chapter 9, namely their tendency to drop off in value quickly. This allows us to approximate a large collection of images, like 5,000, with only a small subset of eigenfaces.

Figure 11.7. A half dozen eigenfaces of the library of twelve U.S. Presidents seen in Figure 11.5.

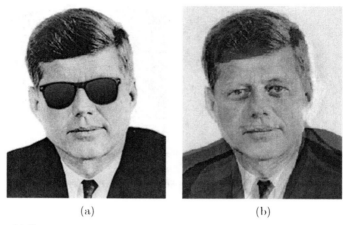

(a)   (b)

Figure 11.8. An altered image (a) of President Kennedy and the image (b) recreated using the six eigenfaces in Figure 11.7.

In the case of our presidential library of images, we have only a dozen images. Nonetheless, let's see how we do with half a dozen eigenfaces seen in Figure 11.7. Specifically, can we recreate the image of President Kennedy found in Figure 11.8 (a), which we see has some minimal disguising?

We subtract the average presidential image from our new image. Then we need to know how much of each eigenface to add for our approximation. We compute the dot product of our mean-subtracted new image and the ith eigenface; let's call this value d_i. Sum the product of d_i and the ith eigenface. Then add the average presidential image. This gives us the image in Figure 11.8 (b). This is part of the power of PCA in facial recognition. One can use a large library of images, reduce it to a much smaller set of eigenfaces, and then use it to recognize a face or even, as we've done here, create an approximation, even with some disguising.

11.4 Recommendation—Seeing Stars

Recommendations are big business, at least for organizations that do them well. But how do they do it? How might *we* do it? "How might Netflix do it?" was the question Netflix asked, when they put together a million dollar competition to improve their recommendation system.

To enter the millionaires' club, you needed to create a computer algorithm that could do a better job of predicting than Netflix's existing recommendation system at the time, called CinematchSM. To create your

Table 11.1. *Ratings of Four People Over Six Movies.*

	Sam	Melody	Mike	Lydia
Movie 1	5	5	0	5
Movie 2	5	0	3	4
Movie 3	3	4	0	3
Movie 4	0	0	5	3
Movie 5	5	4	4	5
Movie 6	5	4	5	5

recommendations, you had Netflix's dataset of users ratings, which are integers from 1 to 5, for all movies. Then, your program would give an anticipated rating for films for a user. This isn't recommending a film, it is supplying a prediction for how a film will be rated by a user.

In the competition, Netflix supplied data on which you could test your ideas. When you thought you had something, you'd test your recommendation method by predicting ratings of movies and users in another dataset that had actual ratings of movies. If your predictions were at least 10% better than the recommendation system Netflix used, Netflix would write a check for a million dollars!

To do our analysis, we'll store the data, not surprisingly, in a matrix. A column is one user's ratings for all the movies with a 0 being a movie that wasn't rated. So, a row contains all the ratings for one movie.

When the Netflix competition was announced, initial work quickly led to improvement over the existing recommendation system. It didn't reach the magic 10% but the strides were impressive, nonetheless. The key was the SVD, which we've seen and used earlier. We saw in Chapter 9 how the SVD can be used in data compression where a lot of data is expressed by less.

So, let's represent our recommendation data in compressed form and use that to do our analysis. Suppose we have Sam, Melody, Mike, and Lydia recommending six movies as seen in Table 11.1.

We store the data in a matrix with six rows and four columns

$$
A = \begin{bmatrix} 5 & 5 & 0 & 5 \\ 5 & 0 & 3 & 4 \\ 3 & 4 & 0 & 3 \\ 0 & 0 & 5 & 3 \\ 5 & 4 & 4 & 5 \\ 5 & 4 & 5 & 5 \end{bmatrix}.
$$

We find the SVD of A where $A = U \Sigma V$. The computation yields

$$U = \begin{bmatrix} -0.4472 & -0.5373 & -0.0064 & -0.5037 & -0.3857 & -0.3298 \\ -0.3586 & 0.2461 & 0.8622 & -0.1458 & 0.0780 & 0.2002 \\ -0.2925 & -0.4033 & -0.2275 & -0.1038 & 0.4360 & 0.7065 \\ -0.2078 & 0.6700 & -0.3951 & -0.5888 & 0.0260 & 0.0667 \\ -0.5099 & 0.0597 & -0.1097 & 0.2869 & 0.5946 & -0.5371 \\ -0.5316 & 0.1887 & -0.1914 & 0.5341 & -0.5485 & 0.2429 \end{bmatrix},$$

$$\Sigma = \begin{bmatrix} 17.7139 & 0 & 0 & 0 \\ 0 & 6.3917 & 0 & 0 \\ 0 & 0 & 3.0980 & 0 \\ 0 & 0 & 0 & 1.3290 \\ 0 & 0 & 0 & 0 \\ 0 & 0 & 0 & 0 \end{bmatrix},$$

$$V = \begin{bmatrix} -0.5710 & -0.4275 & -0.3846 & -0.5859 \\ -0.2228 & -0.5172 & 0.8246 & 0.0532 \\ 0.6749 & -0.6929 & -0.2532 & 0.0140 \\ 0.4109 & 0.2637 & 0.3286 & -0.8085 \end{bmatrix}.$$

We'll reduce to a 2D problem by taking only the first two columns of U, two singular values, and two rows of V. In this way, we can treat the first row of V as coordinates for Sam. So, we can plot Sam's position as the point $\begin{bmatrix} -.5710 & -0.2228 \end{bmatrix}$. We do the same for Melody, Mike, and Lydia.

These points consolidate their movie ratings into two numbers each. This is the same as compressing an image into an image with less storage. We've also made it easier to look at our recommendation system and for the following computations.

Now, who is the most similar? It is Lydia and Sam. Go back, look at the data, and consider the ratings for Lydia and Sam. There is indeed a lot of similarity. The key here is that we never had to look at the data. For large datasets, we can't absorb all the numbers. Further, we use mathematical measures of distance. In this example, we did but also could look at it graphically since we reduced the data to two dimensions. If we had 600 ratings, we would need more than two columns of U to maintain enough detail to reach meaningful conclusions.

Suppose we get a new user, Jess. Suppose her ratings are $\begin{bmatrix} 5 & 5 & 0 & 0 & 0 & 5 \end{bmatrix}$. First, we compress her information to two points. This is done by computing

$$\begin{bmatrix} 5 & 5 & 0 & 0 & 0 & 5 \end{bmatrix} \begin{bmatrix} -0.4472 & -0.5373 \\ -0.3586 & 0.2461 \\ -0.2925 & -0.4033 \\ -0.2078 & 0.6700 \\ -0.5099 & 0.0597 \\ -0.5316 & 0.1887 \end{bmatrix} \begin{bmatrix} 17.7139 & 0 \\ 0 & 6.3917 \end{bmatrix}^{-1}$$

$$= \begin{bmatrix} -0.3775 & -0.0802 \end{bmatrix}.$$

We now have a 2D approximation to Jess's ratings. We can now find to whom she is most similar just as we've done before. If we use cosine similarity, we find she is most similar to Sam. That is, the vector of Jess's 2D data has the smallest the angle with the vector of Sam's 2D data. After Sam, she is closest to Lydia. Now what? This is where various algorithms can be employed and you may want to experiment with your own and do more reading to learn what others have done. For simplicity, we'll note that Jess is closest to Sam and look at what movies Sam has seen and Jess has not, since they, under our measure, are most similar. In particular, Sam has seen movies 3 and 5 and Jess has not. Sam rated movie 5 higher than movie 3 so our recommendation to Jess would be movie 5 and then movie 3.

In this chapter, we've extensively used the SVD for data mining. Many other techniques are used for data mining, with linear algebra being only one field effectively utilized. In the next chapter, we look at rating and ranking, which is another area of analytics. In that chapter, we'll conclude with sports analytics.

12

Who's Number 1?

"Who's number 1?" is a question asked and analyzed by sports announcers before, during, and sometimes even after a season. "We're number one!" is the cheer athletes chant in the pandemonium of success! In this chapter, we learn to apply linear algebra to sports ranking. We compute mathematical "number 1s" by ranking teams. We'll see how the rankings can assist in predicting future play. We'll apply the methods to two tournaments: March Madness and FIFA World Cup. You can learn more about these in [5, 1, 2].

Soon we'll learn how to compute teams' ratings to determine who is number 1 and also who is number 2, 3, 5, 8 and so forth. Here lies the important distinction between ratings and rankings. A *rating* is a value assigned to each team from which we can then form a *ranking*, an ordered list of teams based on a rating method. Solving the linear systems that we will derive produces the ratings. Sorting the ratings in descending order creates a ranked list, from first to last place.

When someone wants to rank teams, a natural rating to start with is winning percentage. This is done by computing each team's ratio of number of games won to the total number of games played. This includes information only about each game's end result and can lead to misleading information. For instance, if Team A only plays one game and they win, then their winning percentage rating would be 1. If Team B plays 20 games and goes undefeated, their winning percentage rating would also be 1. Intuitively, Team B is more impressive since they went undefeated for 20 games, while Team A only won once. But, both teams have the same winning percentage, so this rating system would produce a tie when ranking these two teams. Problems can also arise even if Teams C and D both played 20 games, but C remained undefeated by playing a difficult schedule comprised of the most talented teams in the league while D played only the weakest teams. While D may be a talented team, that fact is not readily apparent from this rating system.

12.1 Getting Massey

A weakness of winning percentage is its indifference to the quality of an opponent. The desire to integrate strength of schedule or the quality of one's opponent into the rating system is why we turn to linear algebra. By computing all the ratings at one time, we can tie strength of schedule into our ratings.

It's interesting to note that the Bowl Championship Series (BCS), which ranked college football teams for the bowl games, used a variation of this method in its ranking methods specifically because they incorporate the strength of an opponent. Kenneth Massey, who has served as a consultant for the BCS, first developed his method as an undergraduate honors math project and it eventually made its way into the BCS [14]. The idea is to use least squares, which was a main topic of Chapter 7.

Here's the idea. Let's use a fictional series of games between teams that play NCAA Division I men's basketball. Suppose Furman University beat Appalachian State University by 1 point. Appalachian State beat the College of Charleston by 15 points. Do we then know that Furman will beat Charleston by 16 points? This scenario is visualized in Figure 12.1 (a), where a directed edge points from the winning team to the losing team and the weight of an edge indicates the difference between the winning and losing scores. Transitivity of scores will rarely, if ever, hold perfectly. Could we assume that it holds approximately? The Massey method does. Let r_1 be the rating for Furman, r_2 the rating for the Appalachian State, and r_3 the rating for Charleston. The Massey method assumes these ratings can predict the outcomes of games. So, $r_1 - r_2 = 1$, since Furman beat Appalachian State by 1 point. Similarly, $r_2 - r_3 = 15$. Let's add an additional game and assume Charleston actually won by 10 points, so $r_3 - r_1 = 10$. This is depicted in Figure 12.1 (b). The scores do not match the transitivity assumption since Furman didn't beat Appalachian State by 16 points. This is often true in sports.

We have three equations $r_1 - r_2 = 1, r_2 - r_3 = 15$, and $r_3 - r_1 = 10$, that correspond to the linear system $M_1 \mathbf{r} = \mathbf{p}_1$, or

$$\begin{bmatrix} 1 & -1 & 0 \\ 0 & 1 & -1 \\ -1 & 0 & 1 \end{bmatrix} \begin{bmatrix} r_1 \\ r_2 \\ r_3 \end{bmatrix} = \begin{bmatrix} 1 \\ 15 \\ 10 \end{bmatrix}.$$

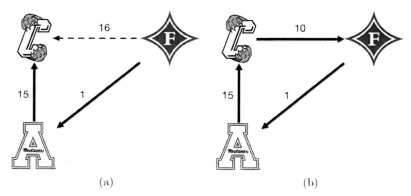

(a)   (b)

Figure 12.1. A fictional season played between three NCAA basketball teams. A directed edge points from the winning team to the losing team. The weight of an edge indicates the difference between the winning and losing scores. In (a), we see a season where transitivity of scores holds. In (b), scores are given that do not match transitivity, which is often the case in a season of data.

Additional games would add more rows to M_1 and $\mathbf{p}_1$. For most sports with a season of data, M_1 will have many more rows than columns.

There are no values for r_1, r_2, and r_3 that will make all three equations simultaneously true. We turn to least squares which leads us to multiply both sides of the equation by the transpose of M_1, in which the rows become columns. So,

$$\begin{bmatrix} 1 & 0 & -1 \\ -1 & 1 & 0 \\ 0 & -1 & 1 \end{bmatrix} \begin{bmatrix} 1 & -1 & 0 \\ 0 & 1 & -1 \\ -1 & 0 & 1 \end{bmatrix} \begin{bmatrix} r_1 \\ r_2 \\ r_3 \end{bmatrix} = \begin{bmatrix} 1 & 0 & -1 \\ -1 & 1 & 0 \\ 0 & -1 & 1 \end{bmatrix} \begin{bmatrix} 1 \\ 15 \\ 10 \end{bmatrix},$$

which becomes

$$\begin{bmatrix} 2 & -1 & -1 \\ -1 & 2 & -1 \\ -1 & -1 & 2 \end{bmatrix} \begin{bmatrix} r_1 \\ r_2 \\ r_3 \end{bmatrix} = \begin{bmatrix} -9 \\ 14 \\ -5 \end{bmatrix}. \qquad (12.1)$$

This system has infinitely many solutions. So, we take one more step and replace the last row of the matrix on the righthand side of the equation with 1s and the last entry in the vector on the righthand side with a 0. This will enforce that the ratings sum to 0. Finding the ratings corresponds to solving the linear system, $M\mathbf{r} = \mathbf{p}$, or

$$
\begin{bmatrix} 2 & -1 & -1 \\ -1 & 2 & -1 \\ 1 & 1 & 1 \end{bmatrix} \begin{bmatrix} r_1 \\ r_2 \\ r_3 \end{bmatrix} = \begin{bmatrix} -9 \\ 14 \\ 0 \end{bmatrix},
$$

which produces the desired ratings. This system gives the ratings $r_1 = -3$, $r_2 = 4.6666$, and $r_3 = -1.6666$. So, the ranking for this group of teams, from first to last, would be Appalachian State, Charleston, and Furman.

12.2 Colley Method

Another ranking method used by the BCS was the Colley method. The Colley method, in contrast to the Massey method, does not use point differentials. So, winning by 40 points counts the same as a win in double overtime by 1 point. Some may argue that all wins are not created equal. However, it can also be that scores create random noise in a system as some games are close until the final minute or a rout is reduced with the second string playing a large portion of the game. The Massey method includes point differentials in calculating a team's rating, so that close games contribute differently from large wins.

The Colley method also uses a linear system, which we'll denote $C\mathbf{r} = \mathbf{b}$, where C is the Colley matrix and $\mathbf{r}$ is the list of ratings. Finding the linear system for the Massey method makes finding the system for Colley method easy. If the linear system for the Massey method, as we found in Equation 12.1, is $M_1 \mathbf{x} = \mathbf{p}_1$ then $C = M_1 + 2I$, where I is the identity matrix. So, C is formed by adding 2 to every diagonal element of M_1. This implies that the vector $\mathbf{b}$ contains information regarding each team's number of wins and losses: $b_i = 1 + \frac{w_i - l_i}{2}$, where w_i and l_i equal the number of wins and losses for team i.

While C and M_1 are associated, it is important to note the derivations of the methods differ. The motivation for the Colley method can be found in [8] or [13]. It isn't surprising that in many cases the algorithms create different rankings for teams over the same season. The ratings will be different as the average of Colley ratings is 1/2 and the sum of Massey ratings is 0.

12.3 Rating Madness

With each method let's rate a small fictional series of games between teams that play NCAA Division I men's basketball. This allows us to walk through

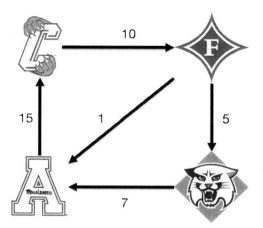

Figure 12.2. A fictional season played between NCAA basketball teams. A directed edge points from the winning team to the losing team. The weight of an edge indicates the difference between the winning and losing scores.

the steps of each method and see their similarities and differences. We'll restrict our discussion to four teams: College of Charleston, Furman University, Davidson College, and Appalachian State University. As in Figure 12.1, we'll represent the records of the teams by a graph where the vertices are teams and a directed edge points from the winning team to the losing team. Although not needed for the Colley method, the weight of an edge indicates the difference between the winning and losing scores. In Figure 12.2, we see, starting in the upper left and moving clockwise, the College of Charleston, Furman, Davidson, and Appalachian State.

For the Colley Method, the linear system is

$$\begin{bmatrix} 4 & -1 & 0 & -1 \\ -1 & 5 & -1 & -1 \\ 0 & -1 & 4 & -1 \\ -1 & -1 & -1 & 5 \end{bmatrix} \begin{bmatrix} C \\ F \\ D \\ A \end{bmatrix} = \begin{bmatrix} 1 \\ 1.5 \\ 1 \\ 0.5 \end{bmatrix},$$

where $C, F, D,$ and A correspond to the ratings for the College of Charleston, Furman, Davidson, and Appalachian State, respectively. So,

$$\begin{bmatrix} C \\ F \\ D \\ A \end{bmatrix} = \begin{bmatrix} 0.5000 \\ 0.5833 \\ 0.5000 \\ 0.4167 \end{bmatrix},$$

which leads to the ranking (from best to worst) of Furman, a tie for second between Charleston and Davidson, and Appalachian State in last place.

If one wishes to integrate scores into the method to differentiate between the teams and break the ties, the Massey method is a natural choice. For the season, depicted in Figure 12.2, the linear system is

$$
\begin{bmatrix}
2 & -1 & 0 & -1 \\
-1 & 3 & -1 & -1 \\
0 & -1 & 2 & -1 \\
1 & 1 & 1 & 1
\end{bmatrix}
\begin{bmatrix}
C \\
F \\
D \\
A
\end{bmatrix}
=
\begin{bmatrix}
-5 \\
-4 \\
2 \\
0
\end{bmatrix}.
$$

Solving, we find

$$
\begin{bmatrix}
C \\
F \\
D \\
A
\end{bmatrix}
=
\begin{bmatrix}
-2.125 \\
-1.000 \\
1.375 \\
1.750
\end{bmatrix},
$$

which leads to the ranking (from best to worst) of Appalachian State, Davidson, Furman, and Charleston. In comparison with the results of the Colley ranking for this collection of games, we see that the Massey ranking rewards for large wins as was the case with Appalachian State over Charleston. It also breaks the tie for second between Charleston and Davidson. An additional piece of information in the Massey method is the predicted score differential. For example, if Charleston and Davidson were to play, the Massey method predicts Davidson would beat Charleston by $1.375 - (-2.125) = 3.5$ points, which while not possible supplies additional analytical information about the teams and their predicted play.

12.4 March MATHness

These ranking methods can aid in predicting the outcomes of future games between teams. Let's see how these ideas can play a role in the NCAA Division I men's basketball tournament known as March Madness. Every March, a large amount of national attention in the United States turns to this single elimination tournament that takes 68 teams from the 350 schools that play men's basketball at the highest level. In the end, a national champion is crowned after three weeks of play. The initial pairings of teams are announced on "Selection Sunday" giving a tournament bracket, a portion of which is shown in Figure 12.3. The matchups shown on the solid lines on the left side of the bracket are the announced games and the madness of March

Figure 12.3. A small portion of the 2014 March Madness bracket. Each vertical line represents a game between two teams, the winner of which is placed on the horizontal line immediately to the right. The matchups on the solid lines at the left are the known games announced at the beginning of the tournament and the teams on the dashed lines show an example of a fan's predictions for the winner of each matchup.

often comes in predicting the winners of each game (dashed lines) to see who advances to eventually win the tournament. By the following Thursday, bracket predictions must be completed and entered in pools in which friends, colleagues, and even strangers compete for pride, office pool winnings, or even thousands of dollars. In fact, in 2014, Warren Buffett insured a billion dollar prize for anyone who could complete a perfect bracket for the tournament. By the end of the second round (or 48 games) none of the over 8 million brackets had perfectly predicted the 2014 tournament. As the tournament progresses, game results either uphold or overturn the predictions of millions of fans. For instance, for the matchups shown in Figure 12.3, Connecticut beat St. Joseph's, which matches the prediction, and Villanova beat Milwaukee, which goes against the prediction. The actual results from 2014 are shown in Figure 12.4.

There seems little madness in applying the Colley and Massey methods to creating brackets for the NCAA basketball tournament. A bracket is formed by assuming a higher ranked team wins. But, this results in everyone who is using the methods to have the same bracket, which may not be desired. An adjustment to the system requires mathematical modeling decisions and results in personalized brackets. To this end, one follows the work introduced in [8] and detailed in [13] and decides a weighting for each game. For

Figure 12.4. A small portion of the 2014 March Madness tournament bracket results.

example, one may decide the recency of a game is predictive of performance in a tournament. One way to measure this is to break the season into n equal parts and assign weights w_i for i from 1 to n. For example, assigning $n = 4$ breaks the season into quarters. Letting $w_1 = 0.25$, $w_2 = 0.5$, $w_3 = 1$, and $w_4 = 1.5$ assumes that a team's play increases in its predictability as the season progresses. For example, games in the first quarter of the season count as 0.25 of a game. Similarly, in the last quarter, games count as 1.5 a win or loss. Games differ in their contribution to the final ratings, increasing the potential of assigning a higher rating to teams that enter the tournament with stronger recent records. Teams that win in predictive parts of the season and do so against strong teams receive a greater reward in their increased rating. The change to the linear systems is minor. Now, a game is counted as the weight of the game in its contribution to the linear systems. Earlier, C and M were formed with each game counting as 1. Now it counts the associated weight. So, returning to our example of breaking a season into quarters, a game would count as 0.5 games in the second quarter of the season. The total number of games becomes the total number of weighted games. The only other difference is the right-hand side of the Massey method. Now, the point differential in a game between teams i and j is a weighted point differential where d_{ij} equals the product of the weight of the game and its point differential. In our example, a game in the first quarter of the season that was won by 6 points would now be recorded as a 6(0.25) or 1.5 point win.

Other aspects of a game can be weighted such as whether a team wins at home or away. Further, one might weight a team's ability to retain winning streaks. Such mathematically-created brackets have been submitted to online tournaments. For example, students at Davidson College who have learned these methods in class have produced brackets that have beaten over 97% (in 2009), 99% (in 2010), and 96% (in 2013) of the over 4 million brackets submitted each year to the ESPN online tournament. Not all brackets perform at this level so constructing one's weightings is part of the art and science of bracketology. One may also wish to compare such methods to a sample of brackets produced with coin flipping or using the ratings produced with winning percentage, as appears in [6].

12.5 Adding Weight to the Madness

Let's return to the NCAA basketball example shown in Figure 12.2 to investigate the effect of incorporating recency into the Colley and Massey calculations. In Figure 12.5 (a), the weights of the edges denote the day of the

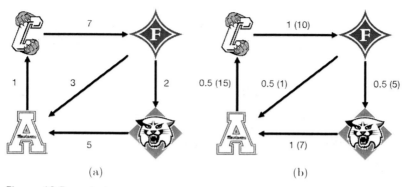

(a)   (b)

Figure 12.5. A fictional season of games played between four NCAA basketball teams. A directed edge points from the winning team to the losing team. The weight of an edge in (a) indicates the day of the seven-day season that the game was played. The same season is given in (b), but now the edges are labeled with the weight for each game and the difference between the winning and losing scores is given in parentheses.

season in which a game was played. Let's take the length of the season to be seven days and weight each game such that games in first half of the season are weighted by 1/2 and in the second half games count as a full game.

The weighted Colley method results in the linear system

$$\begin{bmatrix} 3.5 & -1 & 0 & -0.5 \\ -1 & 4 & -0.5 & -0.5 \\ 0 & -0.5 & 3.5 & -1 \\ -0.5 & -0.5 & -1 & 4 \end{bmatrix} \begin{bmatrix} C \\ F \\ D \\ A \end{bmatrix} = \begin{bmatrix} 1.25 \\ 1 \\ 1.25 \\ 0.5 \end{bmatrix}.$$

Solving, we find

$$\begin{bmatrix} C \\ F \\ D \\ A \end{bmatrix} = \begin{bmatrix} 0.5581 \\ 0.5065 \\ 0.5419 \\ 0.3935 \end{bmatrix},$$

which leads to the ranking (from best to worst) Charleston, Davidson, Furman, and Appalachian State. Incorporating recency also serves to break the tie present in the uniformly weighted Colley calculation.

Now, we turn to Massey for the same series. Figure 12.5 (b) displays the weight for each game (derived from its day played) and the difference between the winning and losing scores is given in parentheses. Now, the linear system becomes

$$
\begin{bmatrix}
1.5 & -1 & 0 & -0.5 \\
-1 & 2 & -0.5 & -0.5 \\
0 & -0.5 & 1.5 & -1 \\
1 & 1 & 1 & 1
\end{bmatrix}
\begin{bmatrix}
C \\ F \\ D \\ A
\end{bmatrix}
=
\begin{bmatrix}
2.5 \\ -7 \\ 4.5 \\ 0
\end{bmatrix},
$$

resulting in the ratings

$$
\begin{bmatrix}
C \\ F \\ D \\ A
\end{bmatrix}
=
\begin{bmatrix}
-0.0476 \\ -2.8095 \\ 2.3810 \\ 0.4762
\end{bmatrix},
$$

and the ranking (from best to worst) Davidson, Appalachian State, Charleston, and Furman. The effect of Appalachian State's large win over Charleston was lessened by weighting recency. Once again, if Charleston and Davidson play a game, the weighted Massey method suggests that Davidson would beat Charleston by $2.3810 - (-0.0476) = 2.4286$ points.

12.6 World Cup Rankings

Massey and Colley rankings were used by the BCS to help rank NCAA football teams. We just saw how to adapt these methods to college basketball. Now, let's implement the rankings to another sport, which already has weightings as part of its official rankings. The FIFA World Ranking is a ranking system for men's national soccer teams. Teams receive points in all FIFA-recognized full international matches. The number of points awarded vary. As we'll see there are several multipliers that combine into the final formula that calculates the points awarded for a game.

The first multiplier correlates to the result of a game. Was it a win, draw, loss, penalty shootout win, or a penalty shootout loss? Table 12.1 lists the multipliers for each of these outcomes. They will be combined with other multipliers to calculate the points a team receives for its performance in a match.

Different types of matches have different importance to teams. This is reflected by the multipliers for types of matches in Table 12.2. A friendly match has a much lower multiplier. These games are essentially exhibition matches, often used to prepare teams for upcoming tournaments or qualifying. The qualifier and World Cup finals games receive much higher

Table 12.1. *Result Points Used in FIFA World Ranking.*

Result	Result Points
Win	3
Draw	1
Loss	0
Penalty shootout win	2
Penalty shootout loss	1

multipliers. However, the importance of a match is not always tied to these types of matches. For example, if two teams are historic rivals, they may take the game quite seriously and their play is similar to tournament play.

Like the bracketology rankings, recency is also weighted. Given the small sample size, matches played over the last four years are included in the calculation. More recent games are given more weight. The multipliers are in Table 12.3.

The FIFA World Rankings also calculates the strength of the opposition and regional strength. All the weightings are combined into the final ranking points for a single match which is

Rank points = 100(result points)(match status)

×(opposition strength)(regional strength),

and is rounded to the nearest whole number. Results for all matches played in the year are averaged, assuming at least five matches were played. The average ranking points for the four previous years, weighted by their multiplier, are added to arrive at the final ranking points.

Now, let's adapt the multipliers in FIFA World Rankings into the Colley and Massey methods. By doing so, the computation of the opposition

Table 12.2. *Multiplier for Different Types of Matches Used in FIFA World Ranking.*

Match Status	Multiplier
Friendly match	1.0
FIFA World Cup and Continental cup qualifier	2.5
Continental cup and Confederations Cup finals	3.0
World Cup finals match	4.0

Table 12.3. *Multiplier for the Recency of a Matches used in FIFA World Ranking.*

Date of Match	Multiplier
Within the last 12 months	1.0
12–24 months ago	0.5
24–36 months ago	0.3
36–48 months ago	0.2

and regional strength are integrated into the ratings produced by the linear system. First, we use FIFA's weightings of games to inform our choices. Then, we calculate ratings for teams that played in the 2014 World Cup tournament.

A weight must be calculated for each game. For March Madness, we based it on time. This can also be done here. However, we'll show how simply using Massey and Colley's measure of strength of schedule is a useful tool. In fact, we won't weight time, even though this could be done and is integrated into FIFA's own rankings. We will add weight to other factors. We will use the same weights for match status. So, if a game occurs in a World Cup final then we will weight it with a 4. We need to include ties. This is easy in Massey as the point differential is zero. However, in Colley decisions must also be made for ties and wins via penalty shootouts. We'll choose to give each team half a win and half a loss for a tie. We will make the same adjustment for penalty shootout wins, viewing them as a tie.

Let's see how predictive of rankings these weights are. There are various ways to test this. We'll take a simple approach and simply see if the higher rated team won when two teams in the World Cup played within six months of the tournament. We will compare this to the FIFA World Rankings at the time of the match. How do we do? We are, indeed, more predictive than FIFA's rankings. In that six month period, FIFA is 66% accurate with 10 right and 5 wrong. There were also seven draws which we don't mark as this work doesn't currently predict draws. Our method, for either Colley or Massey, gets 12 right and 3 wrong during this period and is 80% correct. Could it do better? Can we predict draws? This was a small dataset so is it repeatable and sustainable? If not, would another weighting be better? These are the types of questions that can be explored with this or another ranking.

This research began in the winter of 2014 and is continuing to develop. To share the work, Dr. Mark Kozek of Whittier College, Dr. Michael

Mossinghoff of Davidson College, Davidson College alums Gregory Macnamara and Andrew Liu and I developed the website FIFA Foe Fun (fifafoefun.davidson.edu). The site allows users to make other choices regarding these weightings. Visitors to the site came from over 80 countries before and during the 2014 World Cup.

If you experiment with the site, you'll find that Brazil is often the projected World Cup winner. This agrees with the pre-tournament predictions of many analysts. However, another Davidson College alum, Jordan Wright, used the software and produced a model that perfectly predicted the semi-final and overall winners for the tournament. Jordan's model was submitted to an ESPN online tournament, to which over a million predictions were submitted. His model received more points, given his accuracy, than 98% of the other submissions.

Interested? Develop your own code or try the site and tinker with your own parameters to see who's your number one—in the World Cup, in March Madness, or maybe you're ready for some football or baseball or. . . . What interests you? Find the data and begin developing your model. As we saw, Colley and Massey were adapted from football to March Madness basketball and then to FIFA World Cup soccer. So, think creatively and find your mathematical number one!

13
End of the Line

We've about finished our journey through topics of linear algebra and applications that can be studied with them. As we've learned tools, we've seen a variety of ways to apply our learning.

The early chapters dealt extensively with applications of computer graphics and demonstrated the power and versatility of even simple mathematical operations. Such tools enable the computations to be done quickly, which can be useful and important in video games where quick calculations are necessary for user-defined movement of characters like the Viking in Figure 13.1. For example, Figure 13.1 (a) shows an underlying wireframe model as created for movement while Figure 13.1 (b) removes its lines, enhancing the sense of realism and three dimensional look of the game.

The later chapters delved more into areas of data mining and analytics. In today's world the amount of information available and accessible about our world continually enlarges. Linear algebra is a means to harness this potential power. We've learned some basic methods and many others exist. However, it's important to know that methods continue to be developed, too.

And here is our last lesson. Even the field of linear algebra is far from linear. There are new twists and turns in how methods are used and new techniques that are created. The field cannot be accurately represented by a directed line segment. It continues to grow, encouraged, in part, by the many applications that derive from the usefulness of linear algebra. The boundaries of linear algebra also grow out of the creativity and ingenuity of mathematicians, like you. What applications interest you and came to mind as you read? Do new ideas present themselves as you

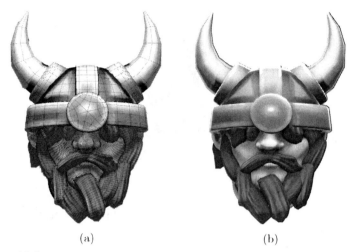

(a)   (b)

Figure 13.1. A wireframe model of a Viking (a) and the character rendered with shading (b).

complete this book? Those ideas can be your steps in the field and possibly take it into new directions. As we've seen, there are many applications and, as you may find as you create your ideas, there are many more to come.

Bibliography

[1] Andrew Beaton. Do it yourself: Predict the world cup, June 2014. [Online; posted June 10, 2014]. 118

[2] Alex Bellos. Belgium to win the world cup? build your own ranking system of the teams playing in brazil, June 2014. [Online; posted June 6, 2014]. 118

[3] Robert Bosch and Adrianne Herman. Continuous line drawings via the traveling salesman problem. *Operations Research Letters*, 32:302–303, 2004. 1

[4] Robert Bosch and Craig S. Kaplan. TSP art. In *Proceedings of Bridges 2005*, pages 303–310, 2005. 1

[5] Elizabeth Bouzarth and Timothy Chartier. March mathness in the linear algebra classroom. *IMAGE*, 52(Spring):6–11, 2014. 118

[6] Tim Chartier. *Math Bytes: Google Bombs, Chocolate-Covered Pi, and Other Cool Bits in Computing*. Princeton University Press, Princeton, 2014. 125

[7] Tim Chartier and Justin Peachey. Reverse engineering college rankings. *Math Horizons*, pages 5–7, November 2014. 60

[8] Timothy Chartier, Erich Kreutzer, Amy Langville, and Kathryn Pedings. Sports ranking with nonuniform weighting. *Journal of Quantitative Analysis in Sports*, 7(3):1–16, 2011. 121, 124

[9] Timothy P. Chartier, Erich Kreutzer, Amy N. Langville, and Kathryn E. Pedings. *Mathematics and Sports*, chapter Bracketology: How can math help?, pages 55–70. Mathematical Association of America, 2010. www.mathaware.org/mam/2010/.

[10] Wesley N. Colley. Colley's bias free college football ranking method: The colley matrix explained, 2002.

[11] G. Dallas. *Principal Component Analysis 4 Dummies: Eigenvectors, Eigenvalues and Dimension Reduction*. georgemdallas.wordpress.com/2013/10/30, accessed September 2014. 76

[12] Victor J. Katz, editor. *A History of Mathematics*. Addison Wesley, 2nd edition, 1998. 3

[13] Amy N. Langville and Carl D. Meyer. *Who's #1:The Science of Rating and Ranking Items*. Princeton University Press, Princeton, 2012. 121, 124

[14] Kenneth Massey. Statistical models applied to the rating of sports teams, 1997. 119

[15] Andrew Schultz. *SWEET Applications of SVD*. palmer.wellesley.edu/ ~aschultz/summer06/math103, accessed September 2014. 88

[16] Tyler Vigen. *Spurious Correlations*. www.tylervigen.com/, accessed September 2014. 66

Index